Letts STUDY GUIDE

AGE 11-14

KEY STAGE 3

SCIENCE

Bob McDuell and Graham Booth

- A clear introduction to the new National Curriculum

- Topic by topic coverage, with lots of diagrams and illustrations

- Frequent questions to test your knowledge

- Index and glossary of terms

- Sample National Test questions and answers

First published 1991
Reprinted 1991, 1992, 1993 (twice), 1994, 1996 (twice), 1997
Revised 1995

Letts Educational
Aldine House
Aldine Place
London W12 8AW
Tel: 0181-740 2266

Text © Bob McDuell and Graham Booth 1995

Illustrations: Ian Foulis Associates, Michael Renouf, Tek-Art, Artistic License
Cover: Veronica Bailey
© BPP (Letts Educational) Ltd 1995

British Library Cataloguing in Publication Data
A CIP record for this book is available from the British Library

ISBN 1 85758 348 5

Acknowledgements
The authors and publishers are grateful to the following for permission to reproduce
photographs and text extracts in this book:

pp24, 74, 107, 138 Hulton Deutsch Collection Ltd; p39 The Harris family; p40
Richard Hutchings; pp41, 42 J L Mason; p54 Science Museum Library; p67 C
Westwood; p67 Simon Fraser; p67 Sinclair Stammers; p68 (Lincoln Cathedral)
RCHME © Crown Copyright; pp68, 77 Alex Bartel; p69 Didier Barrault; p72 Ann
Ronan Picture Library; p72 (Lavoisier) Science and Society Photo Library; p76 Leslie
Garland Picture Library; p78 Cephas Picture Library; p79 Richard Selina (car
exterior); p79 Ford Motor Company (car interior); p81 Network SouthEast; p83 V
Miles; p99 Siemens plc; p143 M Bond; p143 (generator), p145 Science Photo Library;
pp 153-154 Question 1 reproduced by permission of the Midland Examining Group;
Stanley Thornes (Publishers) Ltd for an extract from *Science Companions*, A Porter,
T Wood and M Wood, 1991.

Material from the National Curriculum is Crown copyright and is reproduced by
permission of the Controller of HMSO.

Every effort has been made to trace copyright holders and to obtain their permission
for the use of copyright material. The authors and publishers will gladly receive
information enabling them to rectify any error or omission in subsequent editions.

Printed and bound in Great Britain by
WM Print Limited, West Midlands WS2 9NE
Letts Educational is the trading name of BPP (Letts Educational) Ltd.

Contents

Physical processes

*I*ntroduction

S UCCESSFUL STUDYING AT KEY STAGE 3

During Key Stage 3 of the National Curriculum, you will have to study the following subjects:

English, Mathematics, Science, Technology, a modern foreign language (usually French or German), Geography and History.

If you go to school in Wales, you will also be required to learn Welsh.

This stage of your education is very important because it lays the foundation which you will need in order to embark upon your GCSE courses. The National Curriculum requires you and all 11–14 year olds to follow the same programmes of study, which define the knowledge and skills you will need to learn and develop during your course.

At school, your teachers will be monitoring your progress. At the end of Key Stage 3, your performance will be assessed and you will be given a National Curriculum level. Most students should reach level 5 or level 6, while some may reach levels 7 or 8, or perhaps even higher. In English, Mathematics and Science, you will have to take a National Test towards the end of your last year at Key Stage 3. The results of your tests, also marked in Levels, will be set alongside your teachers' assessment of your work to give an overall picture of how you have done.

How this book will help you

This book is designed for you to use at home to support the work you are doing at school. Think of it as a companion or study guide to help you prepare for classwork, homework, and for the important National Tests. Inside the book, you will find the level descriptions which will be used to assess your performance. We have included them in the book so that, as you near the end of Key Stage 3, you will be able to check how well you are doing.

Also included at the end of the book is a bank of practice questions. These are of the same style and standard as the questions you will face in your National Tests. Attempting these questions in the months leading up to your tests should help you to achieve your best performance.

Reading the book, and doing the questions and activities, will help you get to grips with the most important elements of the National Curriculum in Science. Before you begin to read the book itself, take a few moments to read the following sections on 'Science in the National Curriculum' and 'How to use this book'.

S CIENCE IN THE NATIONAL CURRICULUM

All students in State Schools must study Science as a Core subject up to the age of 16, i.e. the end of Key Stage 4. At the end of Key Stage 4 you will take GCSE in Science (either Double Award, counting as two GCSE subjects, Single Award, counting as one GCSE subject, or the three separate Sciences – Physics, Chemistry and Biology).

Throughout Key Stage 3 and Key Stage 4 you are required to make progress through four different aspects of Science called the four Attainment Targets (ATs). These four ATs are:

- **AT1 Experimental and Investigative Science**
  This is experimental work which you carry out in school. You will find more information about this on page 174. It is not tested in the Key Stage Tests at the end of Key Stage 3.
- **AT2 Life Processes and Living Things**
  This is the area of study which we would call Biology. You will be required to study some of how the human body works, how plants work, the environment, how living things can be grouped and how species change from generation to generation.
- **AT3 Materials and their Properties**
  This is the area of study which we could call Chemistry. It includes classifying materials in different ways, changing materials (including chemical reactions) and grouping materials according to their properties. This section also includes rocks and aspects of Earth Science.
- **AT4 Physical Processes**
  This is the area of study which we could call Physics. It includes electricity and magnetism, forces and motion, light and sound, energy resources and energy transfer and a study of the solar system.

Through each of the three years of Key Stage 3 you should study some aspects of each AT. In May of Year 9 you will take tests which will be set for students throughout England and Wales. These tests will test your understanding of AT2, AT3 and AT4. There are the same number of marks for questions in each of these three ATs.

H OW TO USE THIS BOOK

This book should be used to help you throughout Key Stage 3 to make sure you understand the key facts and issues as you go along. Success in school depends upon regular planned work over a period of time rather than panic bursts of very hard work just before an examination. If you develop the habit of regularly reviewing the work you have done in school, and making sure you understand it, it is less likely to cause you severe problems as the tests approach.

Make an attempt to look regularly at the book and compare what it says with what you have done in school. Make some brief notes – no more than a side of paper on each topic – and keep these to help you to revise closer to the tests. If you have been absent from school, make sure you look especially at the sections you missed.

Throughout the book there are some 'Now Test Yourself' questions. These are not intended to be like the questions in the Key Stage 3 tests at the end of Year 9 – they are there to help you understand the topic.

At the end of the book there is a bank of Key Stage 3 questions. These will be useful as you prepare for tests at the end of Key Stage 3 or for tests in school.

CHAPTER 1

The history of science

You may find it hard to get important scientific discoveries into the correct chronological order. The time chart below shows some of the important scientific advances and when they occurred. You will notice immediately that there has been more progress in the last two hundred years than in any previous time. This reflects the interest in the development of science. It is a fascinating fact that there are more scientists living today than the total number of scientists that have ever lived before.

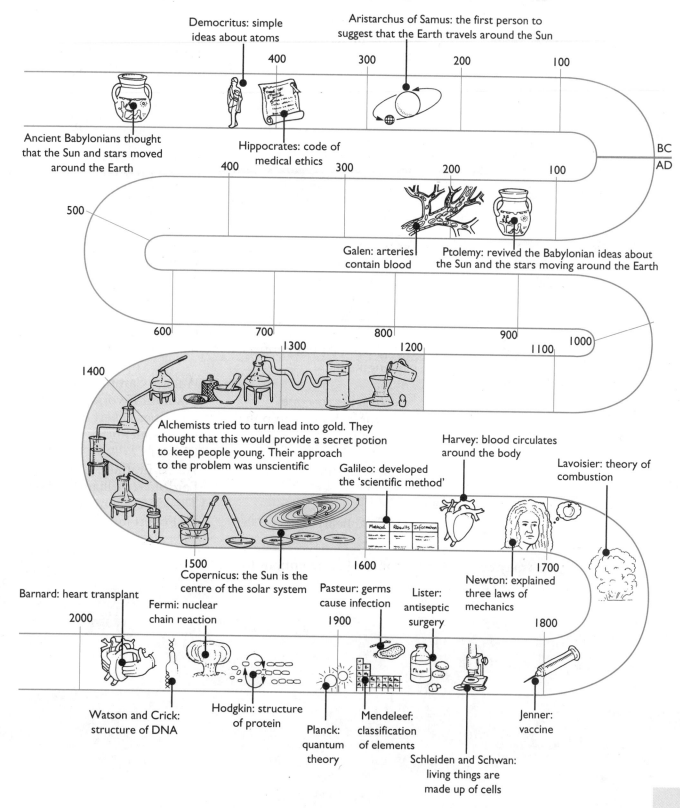

Democritus: simple ideas about atoms

Aristarchus of Samus: the first person to suggest that the Earth travels around the Sun

Ancient Babylonians thought that the Sun and stars moved around the Earth

Hippocrates: code of medical ethics

Galen: arteries contain blood

Ptolemy: revived the Babylonian ideas about the Sun and the stars moving around the Earth

Alchemists tried to turn lead into gold. They thought that this would provide a secret potion to keep people young. Their approach to the problem was unscientific

Harvey: blood circulates around the body

Galileo: developed the 'scientific method'

Lavoisier: theory of combustion

Copernicus: the Sun is the centre of the solar system

Barnard: heart transplant

Fermi: nuclear chain reaction

Pasteur: germs cause infection

Lister: antiseptic surgery

Newton: explained three laws of mechanics

Watson and Crick: structure of DNA

Hodgkin: structure of protein

Planck: quantum theory

Mendeleef: classification of elements

Jenner: vaccine

Schleiden and Schwan: living things are made up of cells

Science time!

You might like to draw a time chart of some of the major scientific and technological inventions of the last four centuries. Your chart could include advances made between the years 1500 and 2000.

The list below will give you some starting ideas but you would need to look up the missing information from reference books. Include your own inventors on your chart. How many famous women can you find?

Date	Invention	Inventor
1590	compound microscope	Zacharias Janssen
1593	thermometer	
	telescope	Hans Lippershey
1643		Evangelista Torricelli
1650	air pump	Otto von Guericke
1714		Gabriel Fahrenheit
1800		Count Alessandro Volta
1822	camera	
	dynamo	Michael Faraday
	Bunsen burner	Robert Bunsen
1866		Alfred Nobel
1870	margarine	Hippolyte Mege-Mouries
1876	telephone	
1885	vacuum flask	
1892	zip fastener	Whitcomb Judson
1895	wireless	
	tape recorder	Valdemar Poulsen
1908	bakelite	Leo Baekeland
1913		Hans Geiger
	tungsten filament lamp	Irving Langmuir
1930	jet engine	
1935	nylon	Wallace Carothers
	electron microscope	Vladimir Zworykin
	electronic digital computer	J. Presper Eckert & John Mauchly
1948		Chester Carlson
	first transistor	Walter Brattain, John Bardeen & William Shockley
1960		Theodore Maiman
1981	artificial heart	Dr Robert Jarvis

You may be able to find some other inventions to add to the list.

CHAPTER 2

Cells

CHARACTERISTICS OF CELLS

All living things are composed of different kinds of **cells**. Each cell is very small and cells build up the organism. The picture below shows typical plant and animal cells as seen through a microscope.

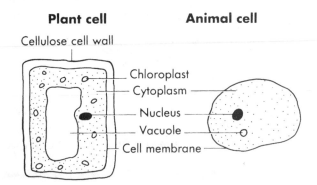

The differences between plant and animal cells are summarized in the table.

Plant cells	Animal cells
chlorophyll present in chloroplasts	no chlorophyll or chloroplasts
cellulose cell walls	no cellulose cell wall
large vacuoles	vacuoles small or absent

Both plant and animal cells contain a **nucleus** that controls the working of the cell. The nucleus contains the genes that are passed on from one generation to the next. Plant cells contain green **chloroplasts** that trap the energy from the Sun and produce food for the plant by photosynthesis. Both plant and animal cells contain **vacuoles**, or fluid-filled spaces. Plant cells contain large permanent vacuoles that are filled with a sugary liquid. Animal cells, if they contain vacuoles, have only temporary ones. Both cells contain a jelly-like substance called **cytoplasm** . Both types of cell have a very thin skin around them called a **cell membrane**. Waste products can pass out of the cell through the membrane and food and oxygen can pass into the cell. Only plant cells have a **cell wall** around the outside that gives the cell a rigid structure. Chloroplasts in a plant cell contain the green chemical **chlorophyll**. This is where photosynthesis takes place.

Amoeba – a single-celled organism

The diagram shows a single cell organism called an **amoeba.** This is found in ditches and stagnant water. Each cell has to carry out all of the jobs of the organism.

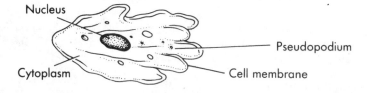

Look at the diagrams which show amoeba. What is happening in A, B and C?

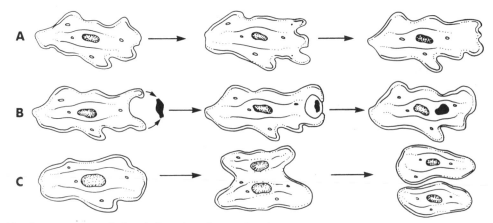

Jennifer has proposed the following hypothesis:

'Amoeba prefer to live in a dark environment rather than a light one.'

How could she attempt to test this hypothesis? Why should she repeat this experiment a number of times?

Cells are usually grouped together to form a **tissue** where they work together. In a tissue all of the cells have a similar job. For example, the leaf of a plant contains a large number of leaf cells whose job is to produce food by photosynthesis. Examples of tissues in our bodies include epidermis (skin), blood, bone and nerves.

Tissues themselves are grouped together in more complex structures called **organs**. Within an organ different cells have different jobs. Examples of plant organs are stems, leaves, roots, flowers and seeds.

H UMAN CELLS

The human body is made up from over 100 billion tiny cells. Unlike an amoeba, different cells in the body have different jobs. All cells, apart from red blood cells, have a nucleus.

The diagram shows some of the specialized cells in the human body and the part of the body where they are found.

A cell in the intestine
Food is taken in or absorbed though the surface of the cell. The folded surface gives a large surface area to absorb food.

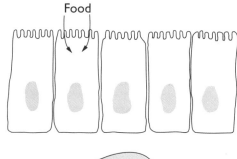

Fat cell
These cells store food as fat.

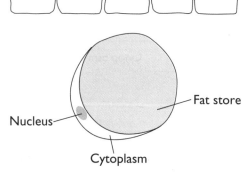

Nerve cell
These cells transport messages around the body.

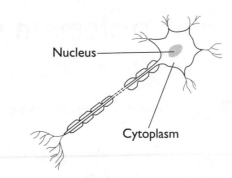

Red blood cell
These cells transport oxygen around the body.

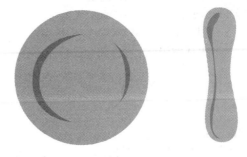

Sperm cell
These cells are sex cells produced by the male. They are very small and have a tail to propel them.

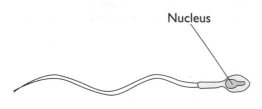

Egg cell or ovum
These cells are sex cells produced by the female. They are much larger than sperm and are unable to move on their own.

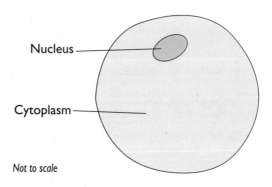

Not to scale

CHAPTER 3

Organ systems in the human body

MAJOR ORGAN SYSTEMS

The organs in the human body work together to form seven main systems in the body.

1. The **circulatory** system
2. The **respiratory** system
3. The **digestive** system
4. The **excretory** system
5. The **skeletal** system
6. The **nervous** system
7. The **reproductive** system

The figure shows the major organs in the human body and the job each organ has in the body.

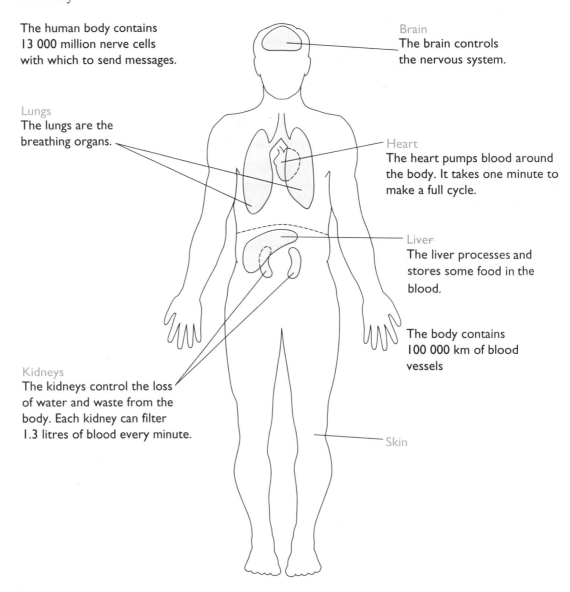

The human body contains 13 000 million nerve cells with which to send messages.

Brain
The brain controls the nervous system.

Lungs
The lungs are the breathing organs.

Heart
The heart pumps blood around the body. It takes one minute to make a full cycle.

Liver
The liver processes and stores some food in the blood.

The body contains 100 000 km of blood vessels

Kidneys
The kidneys control the loss of water and waste from the body. Each kidney can filter 1.3 litres of blood every minute.

Skin

NUTRITION

For good health a balanced diet is essential. The amount of the different types of food that we require varies with our age, occupation and life-style. The main types of food are shown in the table.

Type of food	Benefit to the body	Source
carbohydrates (include sugar, starch)	provide energy	bread, potatoes
proteins	provide amino acids for building and repairing the body	meat, fish, milk, cheese
fats	store energy	butter, oil, margarine
vitamins	required in small amounts for good health, e.g. vitamin C	fruit, vegetables
minerals	required in small amounts for good health, e.g. iron	fruit, green vegetables

In addition to the types of food in the table, water and fibre (roughage) are also required. Fibre (roughage) is not digested but helps in the production of faeces and prevents constipation. There is also evidence that fibre in the diet reduces risks of bowel cancer.

Now test yourself

1 A patient is advised by the doctor to reduce the amount of fat in his diet and to increase the amount of roughage in his diet.

In which form would potatoes be most suitable for him:
chipped potatoes, potato crisps, jacket potatoes, mashed potato, boiled potato?

2 The table gives the constituents of some common foods. Each time, a sample of 100 g of food is taken.
(a) Which foods in the table are a good source of
(i) protein
(ii) carbohydrate?
(b) A diet recommends roast chicken. Why is roast chicken better than a pork chop?

Type of food	Proteins	Fats	Carbohydrates
white bread	8.1	1.7	54.3
brown bread	9.6	3.1	46.7
butter	0.5	81.0	0.0
cheese	25.4	34.5	0.0
milk	3.3	3.8	4.8
roast chicken	24.8	5.4	0.0
pork chop	28.5	24.2	0.0
cabbage	1.7	0.0	2.3
carrots	0.7	0.0	5.4
banana	1.1	0.0	19.2
fruit cake	4.6	15.9	55.0

Discovery of vitamins

Vitamins are complicated chemicals that are required in the body, in small amounts, to control vital processes. Shortage of these vital substances can lead to illness.

We now know the importance of eating fruit and vegetables that provide a source of Vitamin C. Scurvy was a disease that affected sailors on long voyages. The disease caused bleeding gums, weakness of muscles and finally death. It was prevented by giving sailors fruit juice to drink. Dr James Lind found this out in 1747. British ships going on a long journeys took limes to prevent scurvy. This led to British sailors being called 'Limeys'.

The first vitamin to be found was vitamin B_1. In 1896 a Dutch doctor, Dr Eijkman, was looking for a cure for a disease called beri-beri. He noticed that some chickens in the hospital had a similar disease. He then noticed that the condition of the chickens changed when their food was changed from polished rice (rice with the outer husks removed) to whole grain rice (which contained the husks). He and his colleague, Dr Grijns, concluded that it must be a chemical in the husk which prevented beri-beri.

In 1906 they boiled up some husks in water and used the solution to cure a pigeon which was suffering from beri-beri. In 1934, scientists succeeded in identifying the chemical in the husks which cured beri-beri and called it vitamin B_1. In 1937, scientists were able to make vitamin B_1 in the laboratory without having to extract it from rice husks.

So the understanding of the nature and properties of these vital chemicals in the body is fairly recent and is the result of long and painstaking research.

Name of vitamin or mineral	Sources	Use in the body
vitamin A	dairy products, fresh green vegetables, fish oils	good eyesight, growth, healthy skin
vitamin B_1	egg yolks, yeast	good nerves, growth
vitamin B_2	liver, yeast, milk	helps digestion and nerves of skin
vitamin C	oranges and lemons, fresh green vegetables	healthy skin, helps resistance to disease
vitamin D	butter, eggs, fish oils	good bones and teeth
vitamin E	wheatgerm, green vegetables	affects reproduction
calcium	milk, green vegetables	good bones and teeth
iron	meat, spinach, potatoes	good blood
iodine	table salt, seafood	thyroid gland in the neck
sodium chloride	table salt	needed by blood and digestion
potassium	cereals, fresh green vegetables	needed by muscles
magnesium	fresh green vegetables	needed by muscles

Now test yourself

3 (a) Many people today are vegetarians and do not eat meat. Is it possible for somebody whose diet does not contain meat to get all of the essential vitamins and minerals from other sources?

(b) Why is it essential for people, and especially children, to have a regular supply of calcium? Which foods are a good source of calcium?

THE DIGESTIVE SYSTEM

Digestion is the breaking down of large insoluble food molecules into small molecules which the body can use. The digestive process involves:

❶ a mechanical breaking down of the food, e.g. by chewing.

❷ a chemical breaking down using enzymes and acid. Enzymes are biological agents which speed up reactions.

The main chemicals (nutrients) in food are proteins, carbohydrates and fats.

The table summarizes the chemical changes which take place.

Food	Enzymes from	Products
proteins	stomach, pancreas, small intestine	amino acids
carbohydrates	mouth, pancreas, small intestine	sugars
fats	pancreas, small intestine	glycerol, acids

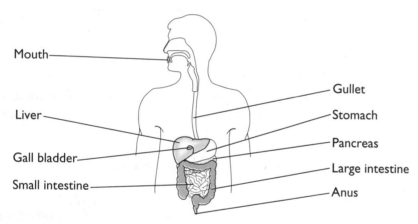

The figure shows the digestive system. Digestion begins in the mouth where mechanical breaking down of the food occurs. The food is mixed with **saliva**, which contains the enzyme amylase. This starts to digest the large starch molecules. The food is swallowed and reaches the stomach. The stomach contains hydrochloric acid and the enzymes which can break down large protein molecules.

The partly digested food is then passed into the small intestine where enzymes from the pancreas continue the digestion. The small molecules produced include glucose and amino acids. These small molecules are absorbed through the lining of the gut into the bloodstream. The blood transports these small molecules around the body.

Undigested food is passed into the large intestine where water is absorbed and faeces are formed. The faeces pass out of the body through the rectum and anus.

Find out the kinds of chemicals that the human body cannot digest.

FOOD ADDITIVES

Food manufacturers put additives in food for various reasons. These include:

1 making it look good
2 keeping it from going bad
3 keeping the ingredients thoroughly mixed
4 reducing the costs of ingredients
5 improving the flavour

Look at food labels on different products to find out their ingredients. For example:

CHERRYADE
Ingredients: Water; sugar; citric acid; flavouring; artificial sweetener (sodium saccharin); preservative (E211); colours (E122, E124).

CHEESE SPREAD
Ingredients: Cheese; skimmed milk powder; butter; whey powder; emulsifying agent (E339); preservative (E202); colour (E160).

Food additives approved for use in the European Community (EC) are given **E numbers,** for example:

E100–E180 are colour agents Many colouring agents are artificial, sometimes even made from coal! Some natural colourings are used (e.g. caramel). You cannot assume that a natural colouring is healthier than an artificial one.

E200–E290 are preservatives They stop microbes growing in the food.

E300–E321 are anti-oxidants They stop air turning the food bad.

E322–E494 are emulsifiers These ensure that the ingredients remain mixed.

Apart from these additives there are sweeteners, flavourings and flavouring enhancers.

An orange squash is advertised as containing no artificial sweeteners. Could sugar be used in the manufacture of this squash?

Food scares?

People are getting very concerned about food additives. Some have been shown to be harmful causing medical problems including allergies.

You might like to carry out a survey in your local supermarket to find out foods which contain additives.

Take care to plan your investigation carefully.

You could ask people questions about whether they look at labels on food or whether they avoid certain additives, etc.

Your survey could lead to some interesting discussions. It is certainly a subject which could affect your health!

Now test yourself

4 The table gives the composition of two different brands of fruit yoghurt per 100 g.

	Brand X	Brand Y
energy value	150 kJ	480 kJ
protein	4.5 g	4.0 g
fat	0.3 g	2.5 g
carbohydrate	5.2 g	18.4 g
additives	preservative, artificial sweetener	preservative

(a) Which brand of yoghurt would be more suitable as part of a slimmer's diet? Explain your reasoning.

(b) Brand Y contains no artificial sweetener. What do you think sweetens this brand of yoghurt?

(c) What method of storage would you recommend if these yoghurts did not contain preservative?

(d) Calculate the mass of protein, fat and carbohydrate in one 125 g tub of Brand Y.

B REATHING AND RESPIRATION

Breathing is the mechanical process of taking fresh air into the lungs and removing stale air from the lungs. Respiration is a chemical process which takes place in cells to release energy trapped in food.

Respiration involves three processes in the human body:

1 oxygen is taken to the muscles by the blood
2 energy is released by the reaction of the food with oxygen
3 carbon dioxide is taken back to be breathed out into the air

There are two sorts of respiration: **aerobic respiration;** and **anaerobic respiration.**

Aerobic respiration requires oxygen. It produces more energy than anaerobic respiration as the food, the fuel, is completely burnt.

glucose + oxygen → carbon dioxide + water + energy

Anaerobic respiration takes place without oxygen and releases less oxygen. It takes place in muscles when there is less oxygen (e.g. in a strenuous race). Lactic acid is produced.

glucose → lactic acid + energy

In a 100 m race, an athlete builds up about 40 g of lactic acid because oxygen is used up faster than the body can take it in.

At the end of the race the athlete has to breathe deeply to replace the oxygen used up.

A marathon runner does not build up an oxygen debt but uses up oxygen at the rate at which it is taken into the body.

The diagram shows the human respiratory system.

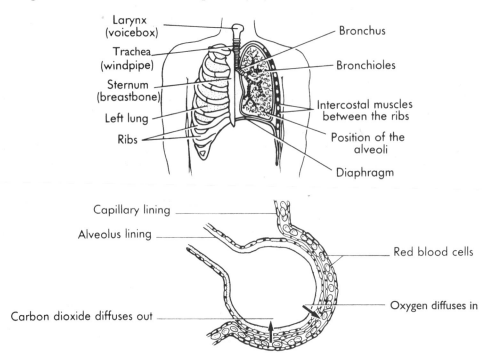

Air enters the lungs, which consist of branched tubes ending in millions of tiny sacs called **alveoli**.

The walls of the alveoli are extremely thin and there is a very large surface area within them. Oxygen can diffuse through the alveoli into the blood from the heart, and carbon dioxide can diffuse from the blood into the alveoli . Rich in oxygen, the blood is then returned to the heart. The blood transports oxygen to the muscles and carbon dioxide back from the muscles.

Exercise makes us breathe more quickly so that more oxygen is drawn into the lungs. Smoking, air pollution and diseases like bronchitis can affect the working of the respiratory system. Find out how!

B LOOD AND ITS CIRCULATION

The human body contains about five litres of a complex mixture called blood. Blood contains:

Red blood cells These carry oxygen around the body from the lungs to tissues throughout the body. The oxygen is carried by the haemoglobin as oxyhaemoglobin.

White blood cells These help to protect the body against disease. There are two types: **phagocytes**, which kill bacteria by engulfing them; and **lymphocytes,** which produce chemicals called **antibodies**. Antibodies are carried around the body killing bacteria and rendering poisons harmless.

Platelets These are fragments of blood cells. They help the blood to clot and so stop bleeding.

Plasma This is a pale yellow liquid which is largely water. It contains dissolved substances such as glucose, amino acids, salts, hormones, blood proteins and waste products.

Label the main parts of blood in the diagram.

Blood flows continuously around the body, pumped by the heart. The blood travels through **blood vessels**. There are three types of blood vessel:

Arteries These carry the blood away from the heart. The blood in the arteries is under pressure and so they have thick muscular walls to withstand this pressure.

Veins These carry the blood back to the heart. The blood is at a much lower pressure and so the walls are much thinner and less muscular. There are valves to prevent the flow of blood in the wrong direction. Veins are usually much closer to the surface of the skin.

Capillaries These are very narrow, thin-walled tubes which join the arteries to the veins. Every organ contains thousands and thousands of capillaries. They have thin walls so that water, oxygen and dissolved foods can diffuse through them to be used by the organ. Waste products can diffuse in the opposite direction.

The figure shows the circulation of the blood in a human.

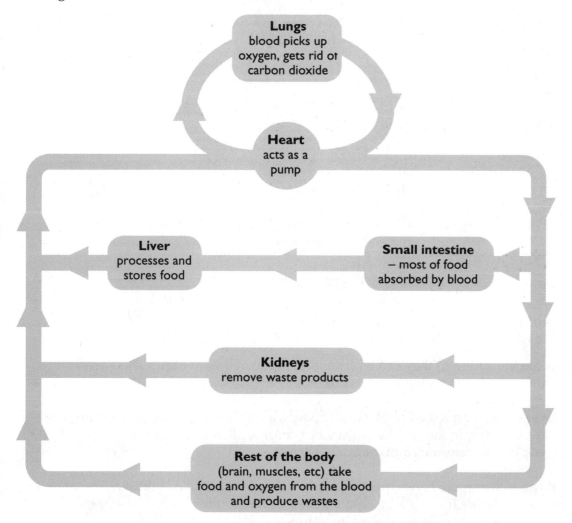

MOVEMENT

There are 206 bones in the human skeleton. The figure shows a skeleton with some of the important bones labelled.

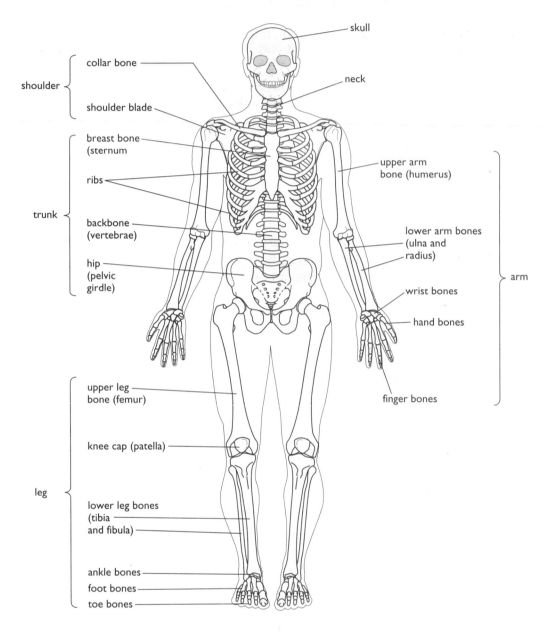

skull

shoulder {
collar bone

shoulder blade

neck

trunk {
breast bone (sternum)

ribs

backbone (vertebrae)

hip (pelvic girdle)

upper arm bone (humerus)

lower arm bones (ulna and radius)

wrist bones

hand bones

arm

finger bones

leg {
upper leg bone (femur)

knee cap (patella)

lower leg bones (tibia and fibula)

ankle bones
foot bones
toe bones

Our skeleton gives the general shape to our body but also protects delicate organs like the heart and the lungs. These are protected by the rib cage.

Bones are strong so they can withstand pressure and hard blows. They are also light in weight to make them easy to move.

To be strong and lightweight, larger bones are made up of hard, dense substance but are hollow in the middle. This hollow part is filled with bone marrow. Red blood cells are produced in the bone marrow.

When two or more bones meet, a **joint** is formed which allows the movement of the skeleton. Bones in a joint are held together by strong fibres called **ligaments**. Movement at joints is controlled by muscles.

Each end of a muscle is connected to a bone by a **tendon** which will not stretch. One end of the muscle is attached to the bone which does not move and the other is fixed to the bone that moves. The muscle pulls on the bone by **contracting**. This means the muscle gets shorter and fatter. When a muscle relaxes it gets longer but cannot exert a pulling force. Another muscle is needed to contract and pull the bone in the opposite direction.

Many bones in the body act as levers which enable a relatively small force to lift a greater object. In the figure below the joint acts as the fulcrum. A small force applied on the contraction of the muscle in the forearm pulls on the bone. The far end of the bone which is carrying the weight is moved through a larger distance.

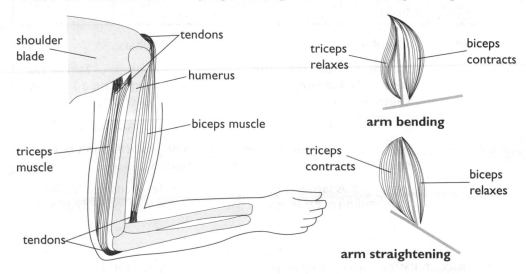

The figure shows two muscles, the triceps and the biceps. These work together to bend and straighten the arm. Muscles all over the body are arranged in pairs in a similar way. Because they work in opposite directions they are known as **antagonistic muscle pairs**.

Most joints in the body allow considerable freedom. These joints are called **synovial joints.** They have features which reduce friction between the bones and allow bones to slide easily over each other:

1. The surfaces of the bones in contact are covered with a smooth layer of elastic cartilage.
2. The joint is enclosed by a tough capsule containing **synovial fluid**, which is a very good lubricant.

There are two types of synovial joint:

Hinge joint (e.g. knee, elbow, wrist)   **Ball and socket joint (e.g. hip)**

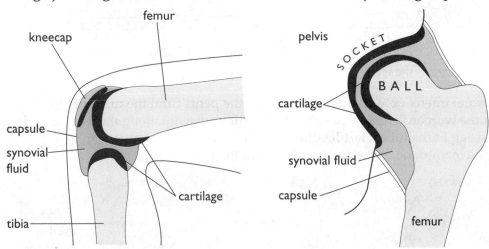

R EPRODUCTION

Human reproduction is by means of sexual reproduction rather than asexual reproduction (page 28).

Changes take place in both boys and girls when they become mature. These changes are summarized in the table.

	Boys	Girls
age of maturity	11–14 years	10–14 years
changes	start to produce male sex cells and a hormone called testosterone	start to produce female sex cells or eggs and a hormone called oestrogen
	voice grows deeper	breasts grow larger
	more hair grows on face and chest	hair grows around pubic region
	penis grows larger and body hair grows round reproductive organs	regular menstrual cycles begin with periods every 28 days

Sexual reproduction usually involves two parents. Both the male and the female produce sex cells. The male and female sex cells were shown on page 11. Male sex cells or sperm are produced by the male in his testes. The female produces female sex cells or eggs in her ovary.

The figure below shows the main parts of the reproductive systems of the male and female.

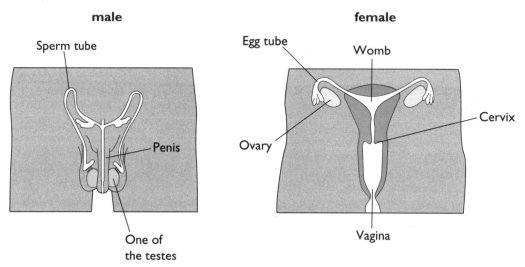

During sexual intercourse sperm pass through the penis from the man into the vagina of the woman. The sperm enter the uterus and swim along the egg tubes. If there is an egg in the tubes, fertilization may take place. Only one sperm breaks through the membrane of the egg and fertilizes the egg.

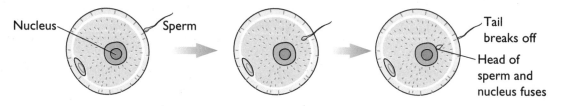

The fertilized egg travels to the uterus and beds itself in the lining. Here the embryo develops. During this time the embryo is provided with the food and oxygen through the umbilical cord. Waste materials from the fetus pass back along the umbilical cord.

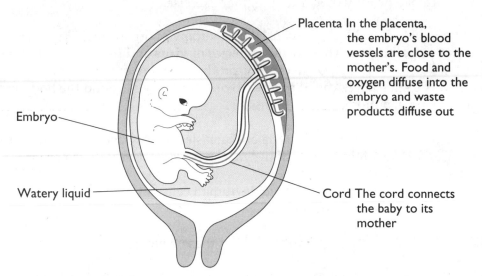

Placenta In the placenta, the embryo's blood vessels are close to the mother's. Food and oxygen diffuse into the embryo and waste products diffuse out

Embryo

Watery liquid

Cord The cord connects the baby to its mother

After nine months of pregnancy, the baby is born through the vagina. A human baby is helpless at birth and relies on care from parents if it is to survive. Some animals produce babies that do not need a great deal of care from parents.

If fertilization does not take place, the uterus lining breaks down and is lost as part of the monthly menstrual cycle.

Identical twins are formed if the fertilized egg divides into two parts and each part develops into a baby. They have the same genes and are either both males or both females.

Non-identical twins are formed when two eggs are released from the ovary and both are fertilized. They do not have the same genes and can be both males, both females or one male and one female.

Now test yourself

5 The table gives information about the number of eggs produced by different species and the care given to the young when they are born.

	No.of eggs per year	Where eggs are fertilized	Care given to egg and young
fish	6 million	in the water	none
frog	3000	in ponds and ditches	none by parents but egg covered with a jelly with an unpleasant taste. No care for tadpoles
bird	6	inside the mother	parents sit on the eggs until hatched, then feed and protect babies
deer	2	inside the mother	mother feeds baby with milk and protects it for 18 months

(a) Suggest another factor which could affect the number of young that mature as adults.

(b) What is the connection between the number of eggs produced and the level of care?

H EALTH

We now know a great deal about how to maintain healthy bodies. We understand the importance of a healthy diet and of regular exercise.

We also know that ill-health can be caused by a variety of organisms. These are called **germs** or **pathogens**. They include species of **virus** and **bacteria**. After they enter the body cells they reproduce rapidly, producing large quantities of poisonous waste substances called **toxins**. Toxins are carried around the body in the blood, producing symptoms of the disease.

Cholera is a disease caused by a bacteria and measles is a disease caused by a virus. Some diseases, e.g. meningitis and influenza, can be caused by bacteria or virus.

Viruses are not able to survive long outside human cells but bacteria can grow and reproduce outside living organisms.

Bacteria can be destroyed using **antibiotics** but these have no effect on viruses.

LOUIS PASTEUR (1822–1895)

In 1765, Lazzaro Spallanzani showed that food would not go bad if the microbes in it were killed. One method of killing the microbes in soup is to boil the soup.

Louis Pasteur was a French scientist who studied microbes and proved that they were responsible for the process of decay. The experiments he carried out are shown below.

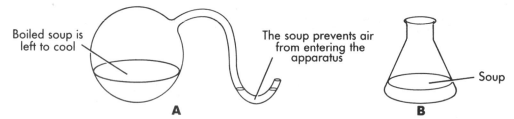

Boiled soup is left to cool

The soup prevents air from entering the apparatus

Soup

A   **B**

The soup in flask A does not go bad. The same soup in flask B goes bad.

Pasteur was able to isolate the germs which caused diseases such as cattle anthrax and chicken cholera.

He was one of the first people to carry out vaccinations to prevent illness.

How can you explain why the soup went bad in B but not in A?

Find out what you can about Edward Jenner and Alexander Fleming.

Effects of smoking

Tobacco contains a drug called nicotine. This drug speeds up the heart rate and raises blood pressure. Because of these changes a smoker has an increased chance of heart disease.

Other problems are caused when tobacco is smoked. The lungs have a mechanism to keep themselves clean. This involves producing a layer of mucus that is moved upward to the throat by moving hairs called cilia. Smoking slows down the movement of the cilia and produces more mucus. This can collect in the bronchioles causing a 'smoker's cough'.

Germs can more easily get into the lungs, making diseases such as bronchitis more common. If this is not cleared up, permanent damage to the lungs can occur.

The bronchioles become narrower and this makes breathing more difficult. The alveoli break up to form large air spaces, reducing the surface area which can absorb oxygen. The person has to breathe faster to receive the same amount of oxygen. This illness, called emphysema, will eventually be fatal.

Smoking can also increase the chances of lung cancer.

Even non-smoking does not remove all of the health risks of tobacco. It has been shown that breathing in smoke from a nearby smoker, known as 'passive smoking', can lead to certain health risks.

Effects of alcohol

Alcohol in small amounts can reduce inhibitions and boost confidence. Alcohol is a depressant drug that slows down the drinker's reactions. Coordination is clumsier and vision can become impaired.

For these reasons there is a severe limit on the concentration of alcohol permitted in a driver's blood. Unfortunately, no driver can know how much alcohol they can drink and still remain 'below the limit' as each human body uses up alcohol at different rates.

The following drinks contain one unit of alcohol:

Half pint of beer   Glass of wine   Single whisky

In large amounts, alcohol causes uncontrollable behaviour. Drinking alcohol can be addictive.

Effects of drugs

Like tobacco and alcohol, other drugs can affect the nervous system which controls the operation of the body. There are four types of drug:

1. **Sedatives** which slow the brain down and make the person sleepy.
2. **Stimulants** which speed up the brain and make the person more alert.
3. **Hallucinogens** cause a person to have experiences which are different from real-life.
4. **Pain-killers** which remove our sense of pain.

Drugs can seriously affect health and cause addiction.

CHAPTER 4

*P*lants as organisms

P ARTS OF A PLANT

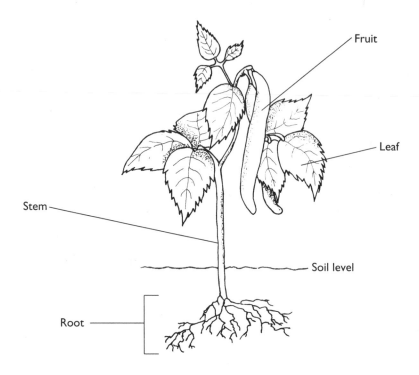

The figure shows the parts of a plant. The roots anchor the plant in the soil. They also take in water and dissolved minerals, especially nitrates, from the soil. The roots include tiny root hairs which are efficient at absorbing water as they have a large surface area. The water and dissolved minerals travel through hollow tubes in the stems to the leaves. The leaves are the 'factories' of the plant. Here the process of photosynthesis produces food for the plant.

Photosynthesis involves the reaction of water and carbon dioxide to produce carbohydrates and oxygen which is returned to the atmosphere. The process only takes place in the presence of sunlight and the green pigment, chlorophyll, which is the catalyst:

carbon dioxide + water $\rightarrow$ carbohydrate + oxygen

Leaves usually have a large area to absorb the maximum amount of light. They are thin so that the carbon dioxide does not have to travel far through the leaf. The veins in a leaf give the leaf some support and provide the leaf with a supply of water. The figure shows a typical leaf.

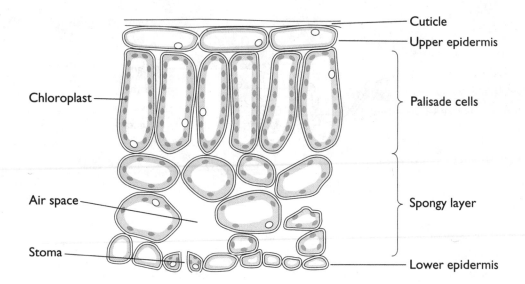

The figure above shows the cross-section of a leaf. Carbon dioxide needed for photosynthesis enters through the **stomata** (singular **stoma**). Most of the stomata are on the underside of the leaf. Stomata open and close. Oxygen and water escape through the stomata. The stomata close at night to prevent too much loss of water.

The waxy layer on the surface of the leaf, called the **cuticle**, prevents evaporation of water from the surface. Below the cuticle there is a single layer of tightly fitting cells called the **epidermis**. The **palisade cells**, below the epidermis but still near to the surface of the leaf, contain a large number of **chloroplasts**. Chloroplasts contain the chlorophyll and it is here that the photosynthesis takes place.

The cells in the **spongy layer**, below the palisade cells, contains irregularly shaped cells and large air gaps exist between the cells. It is here that oxygen and carbon dioxide gases can be stored.

Photosynthesis only occurs in sunlight. The process of **aerobic respiration** occurs in plants all of the time. In sunlight it is not noticed, but out of sunlight the process is more apparent. The equation summarising aerobic respiration is:

carbohydrate + oxygen → carbon dioxide + water

This equation is the reverse of photosynthesis and energy is given out when it occurs.

PARTS OF A FLOWER

Inside a flower there are male and female parts. These enable the plant to reproduce. The male parts are called **stamens** and they make pollen grains inside four-chambered **anthers**. The female parts, called **carpels**, are found at the centre of the flower. They consist of a hollow **ovary** which contains an **ovule**. When this receives pollen the flower starts to form a seed.

Many flowers have brightly coloured **petals** surrounding the carpels and stamens. Sometimes nectar, a sweet, sugary liquid, is produced at the **nectary** at the base of the petals.

Below the flower there is usually an outer ring of green **sepals** which protect the flower when it still just a bud. The figure summarizes the main parts of a flower.

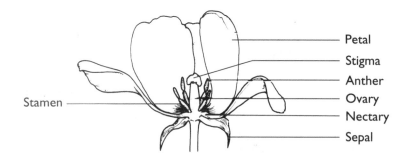

Now test yourself

1

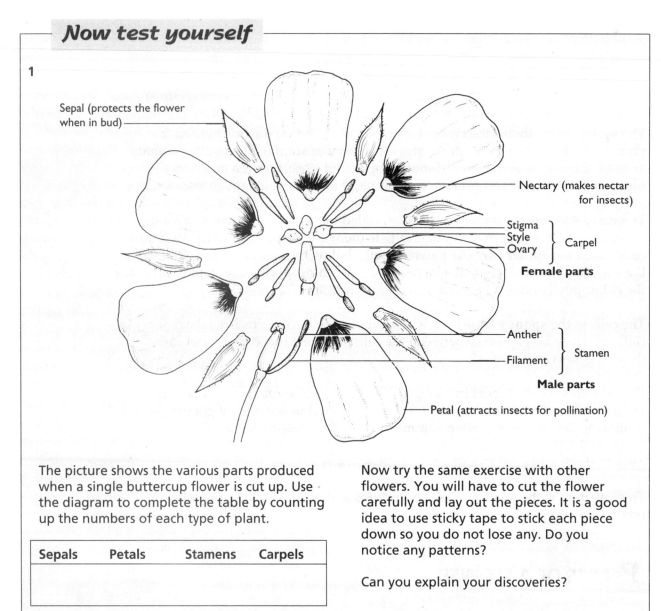

The picture shows the various parts produced when a single buttercup flower is cut up. Use the diagram to complete the table by counting up the numbers of each type of plant.

Sepals	Petals	Stamens	Carpels

Now try the same exercise with other flowers. You will have to cut the flower carefully and lay out the pieces. It is a good idea to use sticky tape to stick each piece down so you do not lose any. Do you notice any patterns?

Can you explain your discoveries?

P RODUCING NEW PLANTS

New plants can be produced in different ways.

Asexual reproduction

Asexual reproduction involves only one parent. It produces offspring which are identical to the parent. Examples of asexual reproduction include:

❶ Single-cell organisms such as amoeba and bacteria. They reproduce by growing and splitting into two identical halves.

❷ Part of a plant can grow into identical plants. Stem cuttings, leaf cuttings, etc. will root and produce plants which are all identical. The illustrations show the steps in producing a new geranium plant by taking a stem cutting.

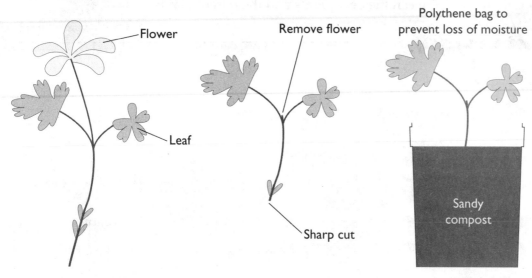

As the offspring are identical to the parent plant, variation does not occur with asexual reproduction. Also any disease, or fault in the parent, will be present in the offspring.

Sexual reproduction

The job of the flower is to produce new plants by sexual reproduction. The first step in this process involves **pollination**. This occurs when a pollen grain lands on its stigma. If the pollen comes from the same flower, **self-pollination** occurs. If the pollen comes from another flower, perhaps carried to the flower by an insect or the wind, **cross-pollination** occurs.

Pollination is followed by **fertilization**, where the male sex cells and the female sex cells join up.

The illustrations show the changes which take place when a flower is fertilized.

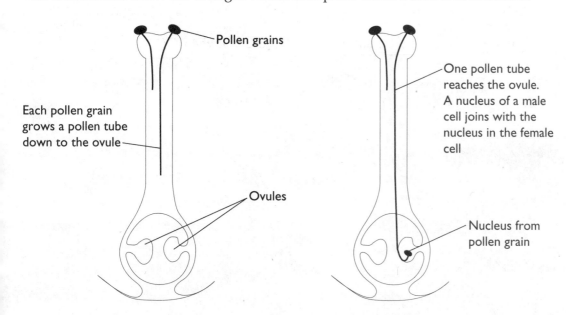

Unlike asexual reproduction, sexual reproduction in plants can produce variation. The seeds produced after fertilization will contain different genetic material and will grow into different plants. Some of these plants will be better in some respects than the parent plants. It is in this way that new varieties of plant are developed.

Plants make fruits to protect the developing seeds and help them to move away from the parent plants. Moving away avoids competition for light, water and plant foods in the soil between the new plants and the parent plants.

Now test yourself

2 Here are the fruits of six plants:

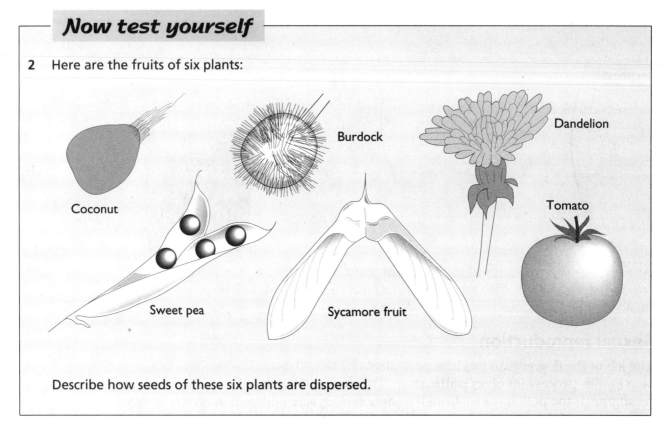

Coconut · Burdock · Dandelion · Sweet pea · Sycamore fruit · Tomato

Describe how seeds of these six plants are dispersed.

G ERMINATION OF SEEDS

The first stage in growing a new plant from seed is **germination**. The following conditions are needed for germination to occur:

1. Water
2. Oxygen
3. Suitable temperature (usually above 5°C and below 45°C)

CHAPTER 5

Classification, variation and inheritance

SIMILARITIES AND DIFFERENCES

This shows the kind of 'spot the difference' puzzle which often appears in magazines. Can you spot the ten differences between the two pictures?

This kind of puzzle requires you to look closely at a detail in one of the pictures and then compare it with the same part of the other picture. Some differences are obvious but others take much more finding.

We often do a similar exercise in science when we compare two similar living things. The figure shows two young forms of different insects – the stonefly and the mayfly. These immature forms are called nymphs.

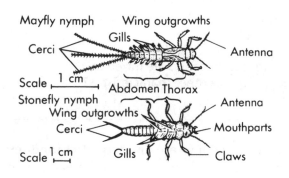

What similarities can you see between these two nymphs?

The table lists differences between these two nymphs.

Stonefly nymph	Mayfly nymph
long antennae	short antennae
large body	small body
projecting mouthparts	no mouthparts visible
large head	small head
distinct neck	no neck
three visible segments in thorax	two visible segments in thorax
wing outgrowth on two segments	one pair of wing outgrowths
gills on thorax	gills on abdomen
pair of claws on each leg	single claw on each leg
no hairs on first part of first leg	hairs all along first leg
two cerci	three cerci
ten abdominal segments	eight abdominal segments

Now test yourself

1 Look at the two organisms A and B. List three differences between them.
2 Similarly, list three differences between C and D.

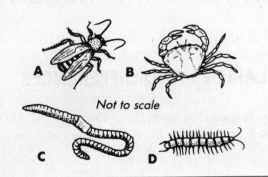

Not to scale

DIVERSITY OF ORGANISMS

Biology is the study of living things. The word comes from a Greek word, *bios*, meaning 'life'. It is easy to identify distinguish a living thing from a non-living thing. For example, a living thing can grow and a non-living thing cannot.

Now test yourself

3 Growth is only one of the characteristics of a living thing.

Here are six more, but the letters are mixed up. Can you work out the six other characteristics?

(a) VOEMMTEN (b) POSSERNEC
(c) ORDERTINCUPO
(d) ITREECXNO (e) GDEFEIN (f) GEYREN DENE

Now try and write a sentence about each one, explaining what it means.

All living things are divided into **kingdoms**. Two of these are the animal kingdom and the plant kingdom. Kingdoms are then divided into **phyla** (singular **phylum**).

There are millions of different **species** of plants and animals and new species are being discovered all of the time. Rather than try and study each one separately, it is sensible to put them together and study them as groups.

For example, if we consider the animal kingdom, we can divide all animals into two major groups: animals with backbones and animals without backbones.

Animals with backbones are called **vertebrates** and animals without backbones are called **invertebrates**.

Now test yourself

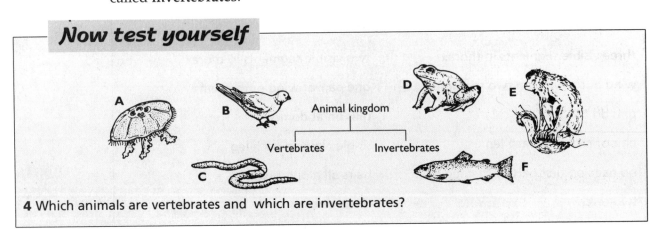

4 Which animals are vertebrates and which are invertebrates?

We can divide vertebrates further according to whether the animal has a body temperature the same as its surroundings (called **cold blooded**) or whether the animal is able to keep its body temperature the same even on hot or cold days (**called warm blooded**).

Now test yourself

5 Which animals G to L are warm blooded and which are cold blooded? Remember a cold-blooded animal cannot control its body temperature. Its body temperature is the same as the temperature of its surroundings. Warm-blooded animals control their body temperature and keep it at a constant level.

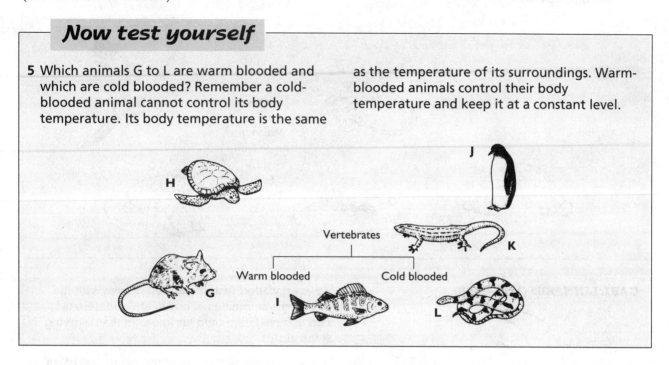

Cold-blooded animals cannot use internal processes to control their body temperature. However, a snake can warm up its body by basking in the sun.

The table lists common groups of vertebrates divided into cold-blooded and warm-blooded **classes**.

cold blooded	warm blooded
fish	birds
amphibians	mammals
reptiles	

Each of these classes have their own characteristics.

Fish	Paired fins, gills.
Amphibians	Slimy skin. Spend some of lives in water.
Reptiles	Dry scaly skin. Lay eggs on land.
Birds	Feathers. Lay eggs on land.
Mammals	Hair. Provide milk for young from special glands.

Now test yourself

6 Look at each of the illustrations and decide which one of these five classes each animal belongs to.

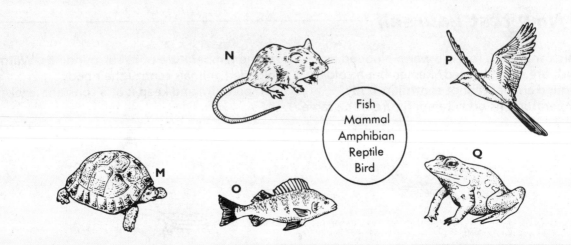

Fish
Mammal
Amphibian
Reptile
Bird

CARL LINNAEUS (1707–1778)

Carl Linnaeus was born in Sweden in 1707. He was the son of a church minister. He went to medical school to train as a doctor, but his main interest was in plants. He spent much of his spare time searching the countryside for new plants and then identifying them. He visited Lapland and wrote a book about the plants there.

He even visited Britain and fell in love with the yellow gorse bushes of our heathlands. He is said to have fallen onto his knees in thanksgiving at the sight!

Linnaeus realized that the same name was often given to more than one plant. Sometimes a plant would be given different names in different parts of the same country. He felt that every plant or animal should have its own individual scientific name which people throughout the world would recognize.

He used the ideas of the English botanist John Ray, who had already started using Latin to name plants. The naming system he used consists of two words. The first word gives the **genus** or family) which the organism belongs to and the second word gives the **species**. For example, the columbine which flowers in our gardens in May is called *Aquilegia hybrida*. *Aquilegia* is the genus name and *hybrida* is the species.

MARIANNE NORTH (1830–1890)

Marianne North was one on the most famous women botanists in the 19th century. Botany, the study of flowers and plants, has interested a number of women scientists.

Marianne visited the Royal Botanical Gardens at Kew where she drew plants and studied their behaviour. She travelled through Europe and the Middle East, making detailed records of the plants she saw. After the death of her father in 1869, she decided to visit different parts of the world to see plants growing in their natural environments. She travelled to Jamaica, North and South America, Japan, India, Australia, South Africa and the Seychelles. Of course, travelling was not as quick and easy as it is today. As she travelled, she discovered four new plant varieties and brought many new plant species back to England. A special art gallery was built at Kew to display her paintings.

She never married and used money left by her father on his death to support her travels and work. Her married sister, Catherine, also painted flowers. However, she did not have the opportunity to travel and her work today is largely unknown.

We can classify invertebrates in a similar way producing a number of different groups or phyla.

Phylum	Feature
Protozoa	made of one cell, e.g. amoeba
Sponges	animals made of cells loosely joined together
Cnidaria	body walls made of two layers of cells, e.g. jellyfish, sea anemones
Flat worms	flattened worm-like shape
Annelida	worms made of segments, e.g. earthworms
Arthropoda	jointed legs, bodies made of segments; includes spiders, insects, centipedes
Mollusca	no segments; a fleshy pad on which they crawl, e.g. slug, snail
Echinodermata	star-shaped pattern – spiny skin, e.g. starfish

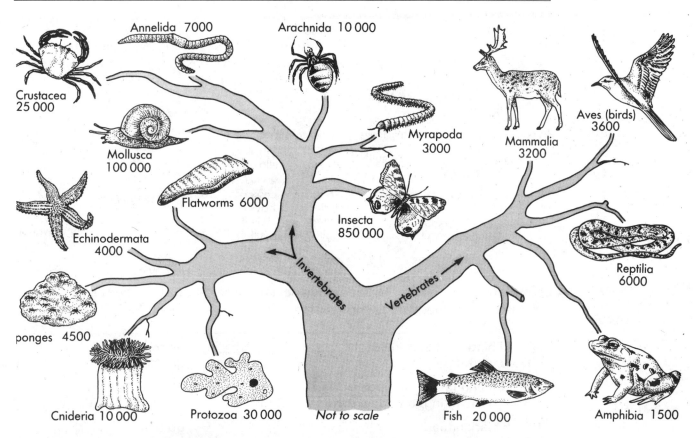

The figure above summarizes the family tree of animals. The number in each case is an approximation of the number of different species.

Do you realize that if you went out into the rain forest with a net looking for insects and spiders, there is a very good chance that you would find a species which has never been found before? It is believed that at least three-quarters of tiny creatures like insects and spiders have never been positively identified. It is possible that a new species could be found every day.

In the same way, we can classify plants into different groups. The plant kingdom is divided as follows:

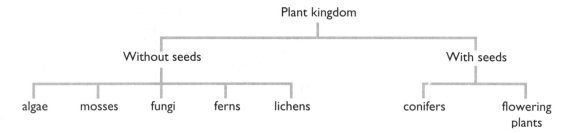

KEYS

Having seen the wide diversity of living things and ways in which we group them, we are going to see how we can identify a living thing by using a **key**. A key is a series of questions which leads to the identification of individual plants or animals. At each point in the key, the question divides the organisms into two groups depending upon whether or not they have a certain characteristic.

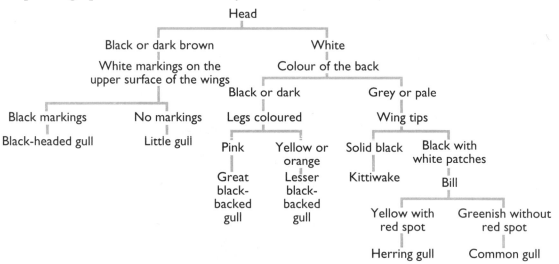

Look at the figure above. This identifies seven species of gull. The key can be expressed in another way as shown in the table.

1 Head black or dark brown	go to **2**
Head white	go to **3**
2 Black markings on the upper surface of wings	Black-headed gull
No black markings on upper surface of wings	Little gull
3 Back black or dark	go to **4**
Back grey or pale	go to **5**
4 Legs pink coloured	Great black-backed gull
Legs yellow or orange	Lesser black-backed gull
5 Wing tips solid black	Kittiwake
Wing tips black with white patches	go to **6**
6 Thick yellow bill with red spot	Herring gull
Thin greenish gill without red spot	Common gull

Now use the key to identify the three gulls on the next page.

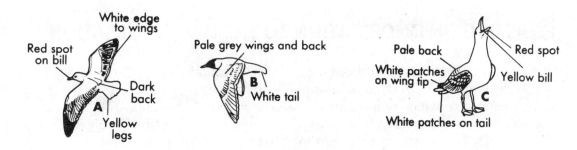

If you follow through the key systematically, i.e. one step at a time starting at the beginning, you should find that **A** is the lesser black-backed gull (head white, dark back, yellow legs), **B** is the little gull (black head, no black markings on the upper surface of the wings) and **C** is the herring gull (white head, pale back, white patches on wing tip).

Now test yourself

7 Use the key to identify the six leaves in the illustration.

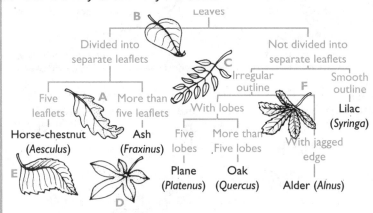

8 Make your own key to identify each of the fish shown here.

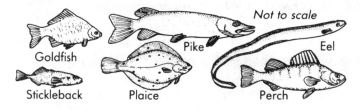

You could start by asking questions such as: *Has the fish a tail fin?*

The eel has no tail fin. The other five fish have tail fins.

Now carry on. There are many different ways of doing it.

9 The picture shows four arthropods found in a wood. Use the key to identify them.

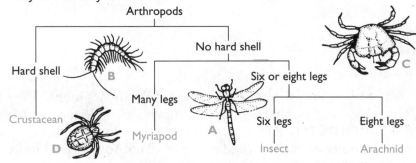

PASSING ON INFORMATION TO THE NEXT GENERATION

All the cells in our bodies, apart from red blood cells, contain nuclei. These nuclei contain **chromosomes** which store information to enable new similar cells to be produced and information which can be passed on to future generations. The genetic information is stored on threads of **DNA**. Every human cell contains 23 pairs of chromosomes. Twenty-two pairs are ordinary pairs. The other pair are the sex chromosomes which determine the sex of the person. If the sex chromosomes are alike (called XX) the sex is female, and if they are different (called XY) the sex is male.

Sexual reproduction involves special cells called **sex cells**. When cells divide, during growth or repair, they produce identical cells. This process is called **mitosis**. After the division, each cell contains the same number of chromosomes as the parent cell.

The figure shows mitosis taking place in a cell with only four chromosomes.

The rest of the cell divides to form two new, identical cells

Two pairs of chromosomes

The chromosomes double

The chromosomes separate into two groups

The nucleus divides into two new nuclei, each with four chromosomes

Sex cells are not produced by mitosis but by a process called **meiosis**. This time the chromosomes make an exact copy of themselves but the parent cell divides into four new cells. Each new cell has half the number of chromosomes of the original cell. Here is how this happens, again with a cell containing four chromosomes.

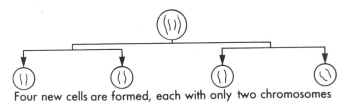

Four new cells are formed, each with only two chromosomes

The male and female sex cells join together during **fertilization** to produce a new cell (the **zygote**) which has characteristics of both parents and develops into the **embryo**. This is summarized in the figure below.

Female gamete (two chromosomes)   Male gamete (two chromosomes)

Fuse

Zygote   (four chromosomes)

The instructions carried by a chromosome for a particular characteristic, such as eye colour or blood group, are called **genes**. Every human being has two copies of each gene in every normal body cell, one in each chromosome. One gene comes from the father and one from the mother.

VARIATION

Have you noticed that brothers and sisters in the same family are not identical? Unless there are identical twins, the chances of parents having two identical children are about one in 1 800 000 000 000 000. This number is very large. In fact, this number is much larger than the number of people who have ever lived. You will not meet your exact double during your lifetime.

Variation within a family is caused by new genes formed by mutation and different mixes of genes. There are two types of variation:

➊ **Discontinuous variation** This enables us to separate the population into different clearly distinguished groups, e.g. by blood group. We can sort the blood groups of individuals into four main groups: A, B, AB and O. No one falls between two groups (e.g. a mixture of groups A and O).

Another example of discontinuous variation is albinism. This is a complete lack of skin pigment caused by the difference of a single gene. An albino will never become tanned however much sun they are exposed to.

➋ **Continuous variation** Sometimes we cannot see clearly different groups.

For example, if we were to measure the length of the middle finger of the right hand of thirty children, the results could not be clearly put into groups. Height and weight are other good examples. The graph shows the kind of variation which could be seen in the heights of a sample of men.

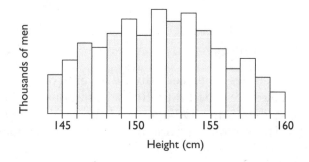

Whether the variation in height of these men is due to genes inherited from parents, or whether it depends upon the way the men have lived, is a question which has interested scientists for many years.

In order to investigate this, scientists have studied identical twins. Identical twins have exactly the same genes because they are both formed from a single fertilized egg. This fertilized egg splits into two after it has been fertilized and two identical embryos are formed. So any differences in identical twins cannot be due to differences in their genes. Any differences must be due to the influence of the environment, such as amount and quality of food. Studying identical twins who are separated shortly after birth and reared separately can give interesting information. If the twins have similar characteristics, it suggests that inheritance was the main influence.

Now test yourself

10 A study was carried out on a hundred pairs of identical twins. For fifty pairs of twins, each pair was brought up together. For the other fifty, the pairs were separated at birth and brought up apart. The results are summarized in the the table below.

Difference in characteristic	Twins brought up together	Twins brought up apart
height (cm)	1.6	1.7
weight (kg)	2.0	4.8

Look at the information in the table. Compare the differences in heights of twins brought up together and separately. Is there much difference? Now, compare the differences in weights of twins brought up together and separately. Is there much difference this time? Can you draw any conclusions from this information?

11 Here are some beetles found living together. You will notice they are not all the same. In the box you will see how we could measure the length of a beetle.

(a) What features do all the beetles have in common?
(b) What features about the beetles vary?
(c) Which features show continuous variation and which discontinuous variation?
(d) Display your results in tables, bar charts or other ways which you think will show the results more clearly.

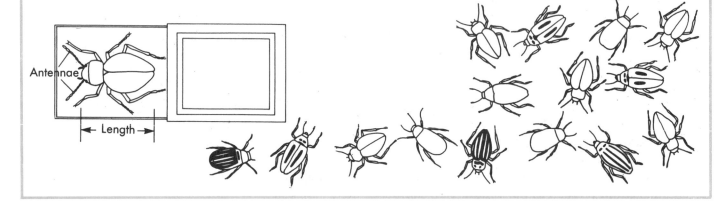

MUTATIONS

The copying of chromosomes when cells divide is very complicated and mistakes can occur. These mistakes are called **mutations**. Mutations can be caused by radiation and by some chemicals. Radiation and chemicals which cause mutations are called **mutagens**. These mutagens are found in certain drugs, cigarette smoke and the fumes from certain types of plastics when burning.

Down's syndrome is caused by a mutation. Children suffering from Down's syndrome have an extra chromosome. This occurs most frequently when the mother is older and cell division to produce eggs has not occurred properly.

NATURAL SELECTION

Charles Darwin proposed the idea of natural selection after observing many different plants and animals. He was a young scientist when HMS Beagle set sail on a five-year voyage on December 27th 1831. His observations on this voyage led him to propose a theory to explain how evolution had occurred. This was the theory of **natural selection.**

This theory suggested that organisms were able to produce large numbers of offspring. Some of these will never reach maturity. The organisms which survive are the ones which are best adapted for the environment. Over a number of generations the population consists of the organisms which are best adapted to the environment.

An example of natural selection is shown by the peppered moth. The peppering effect provides good camouflage for the moth. Moths are eaten by birds. Light-coloured moths on light backgrounds have a better chance of survival than dark moths on light backgrounds. The light-coloured moths survive to breed and so light moths predominate.

The peppered moth has a speckled grey colour. Before 1850 all peppered moths were a light-grey colour. The black form was very rare. By the end of the 19th century, 95 per cent of all the moths in industrial cities were black but in country areas the grey moths were still common. The black form had an advantage in city areas where they were camouflaged against blackened buildings. In country areas the black moth had no advantage.

S ELECTIVE BREEDING

Throughout history, humans have attempted to improve plants and animals by artificial selection. This is also called **selective breeding**. The scientist examines the natural population and selects individuals with particularly desirable characteristics. These individuals are then bred together. This is repeated through several generations.

In growing cereals, a scientist might choose varieties which have sturdy stems and a large number of seeds per stem to produce a better variety for cultivation.

Chapter 6

*L*iving things in the environment

THE RHYTHMS OF LIFE

Living things respond to seasonal and daily changes in different ways. Here are some examples of this:

1. A deciduous tree (e.g. oak) sheds its leaves in autumn and grows new leaves in spring. The pigments in the leaves change at different times of the year.
2. A tortoise is an example of an animal which hibernates during the winter. The tortoise's metabolism slows down during the winter enabling it to conserve energy and live through harsh conditions. It uses up a store of energy during hibernation. In spring the tortoise becomes active again.
3. A chrysanthemum comes into flower in the autumn when the length of the day is decreasing. Nurserymen, wanting to get chrysanthemums in flower at other times of the year, have to reduce artificially the amount of light each plant receives.
4. Nocturnal animals become active at night. In a zoo, nocturnal animals are kept in the dark so that visitors can see them moving about.

Can you think of any other examples?

HABITATS

A place where a community of organisms live is called a **habitat**. The drawing shows some plants and animals which live in a wood. This kind of environment is called a terrestrial habitat. Conditions in the wood are often cool and damp with limited amounts of light reaching the ground.

The picture on page 44 shows plants and animals which live in a pond. This is a very different environment from a wood, so different plants and animals will live there. A pond is an example of an aquatic habitat. Every organism lives in a habitat where it can survive successfully. How can we tell if a species is surviving well in a habitat? What factors affect the number of creatures that live in a habitat?

Not to scale

Some animals move away from light (e.g. woodlice, earthworms and cockroaches). This is because light is associated with warmth which would cause such animals to lose moisture, dry out and die. They tend to thrive in cool, damp conditions (e.g. under stones).

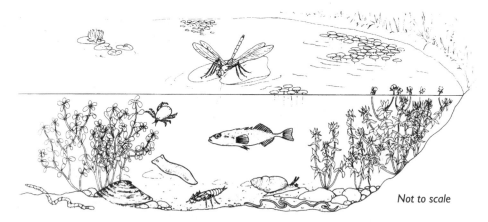

Not to scale

Trees in a forest grow tall to reach the light which is necessary for making food by photosynthesis and for growth. Mosses and ferns are happier living in the shade of the trees in the forest. There it is cooler and there is less chance of drying out.

Now test yourself

1 The two nymphs on page 31 live in fast, flowing rivers. They crawl around underneath stones. Why are they suitable for survival in this habitat?

2 The table shows animals collected in two sites:
Site A, a damp, shaded area of the garden;
Site B, a dry open pasture with a sunny position.

Animal	Number of animals found at location	
	A	B
snails	55	1
worms	20	5
centipedes	5	1
ants	40	25
spiders	20	15
beetles	20	10
aphids	20	40

(a) Which animal occurred in larger numbers at site B rather than site A?
(b) Complete the pie chart opposite, showing the distribution of animals at site A. Each segment is 20°.
(c) Which animal is likely to be a secondary consumer? Give a reason for your answer.
(d) Suggest **two** reasons why there are more snails at site A than at site B.

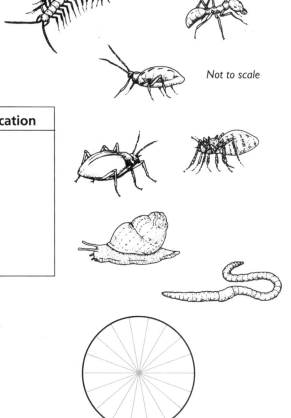

Not to scale

FOSSILS

Fossils are traces of organisms preserved in rocks. Scientists expect such rocks to be older than a few tens of thousands of years to qualify as a fossil. Most fossils were formed when the remains of plants and animals were buried by sediment. Only the hard part of the organism will remain (e.g. the shell or an impression of the shell in a rock). The scientist has to decide what the original creature was like by looking at similar organisms living today.

The types of fossil found in a particular area give some idea of the organisms which existed in that location. The picture below shows some fossils found in the Peak District, in Derbyshire. All of these fossils are remains of sea creatures.

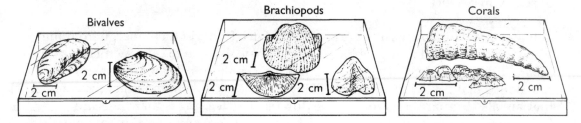

Scientists therefore know that Derbyshire was under the sea when these animals were alive about 300 million years ago.

Making your own fossil

This investigation will help you to see what took place over millions of years. Take a shell and press it into a piece of Plasticine. Make sure you have a clear imprint of the shell.

Now mix some plaster of Paris with water until you have a very thick, smooth paste. Pour the plaster into the Plasticine mould and leave the plaster to set. Carefully remove the Plasticine and you will have a plaster copy of the original shell. Fossils formed in the earth in a similar way with other rocks taking the role of the plaster.

How would you guess the age of a real fossil? Find out how scientists discover this

F EEDING RELATIONSHIPS

Food is important to all of us; it is the source of the energy that our bodies need to keep them working and enable us to move. It also gives us small quantities of important chemicals such as iron and vitamin C that our bodies need to work properly and the proteins that we need for growth.

Green plants are the world's food factories. They range from the microscopic plants in sea plankton to giant oak trees. When the conditions are right these plants are all busy making food.

Because green plants are the only things that make food, they are called **producers**. Using energy from the Sun, the plant makes food in the form of **glucose**, a form of sugar. The glucose is made from **carbon dioxide** and **water**.

Once the plant has made glucose, it can convert it into other carbohydrates such as **starch**. Starch is the form in which plants store food. Potato plants store starch underground in tubers; other plants may store starch in the leaf, like cabbage, or in the stem, like celery.

You may not like cabbage, but slugs certainly do. They get their energy from the carbohydrates stored in the cabbage leaves. They also like to burrow their way into potato tubers that are underground. Slugs are examples of **primary consumers,** the first things that eat the food produced by the plant. When humans eat the cabbage they are also primary consumers. The difference in feeding is that slugs are **herbivores**: they only eat plants. Many humans are **omnivores**: they eat animals as well as plants.

Now test yourself

3 Blue tits eat the aphids that feed on oak leaves. The blue tits are eaten by kestrels.

Draw a food chain that shows these feeding relationships.

4 A food chain involving some organisms in the sea is shown below:

plant plankton → animal plankton → sand eel → salmon → seal

(a) Which organism is the producer for the food chain?

(b) Name the secondary consumer in this chain.

(c) Which organism is likely to be a herbivore?

(d) Explain why you cannot tell from this chain whether the sand eel is a carnivore or an omnivore.

(e) What is the energy source for the food chain?

A **food chain** shows how energy passes from the producer to the consumers. When a human eats cabbage the food chain looks like this:

cabbage → human

This is a very simple food chain: only two organisms are involved.

A food chain involving the cabbage and the slug can be more complex. The slug eats the cabbage and in turn is eaten by a small bird such as a thrush. This is then eaten by a kestrel. The kestrel is a **carnivore**; it only eats other animals. The thrush is an omnivore; like humans, it eats both animals and plants.

The food chain for the cabbage, slug, thrush and kestrel looks like this:

cabbage → slug → thrush → kestrel

In this food chain, the thrush is called a **secondary consumer** and the kestrel is a **tertiary consumer**; these are the second and third consumers in the chain.

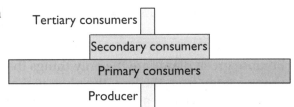

Many animals eat a variety of things and in turn are eaten by a variety of things. Humans eat a wide variety of producers, as well as primary and secondary consumers. An oak tree can support a large number of food chains, with several organisms appearing in more than one chain. In a habitat such as a tree, or a pond, the food chains can be put together to form a **food web**. The diagram shows the food web based on a single woodland tree.

This food web contains a number of food chains; some organisms, such as beetles, are only in one food chain while others, such as swallows, are in several.

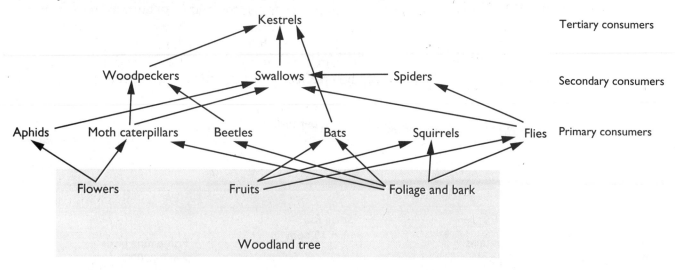

Now test yourself

5 (a) Write down the food chain that involves beetles.

(b) How many separate food chains involve swallows?

(c) Which primary consumer does not have a predator shown in the food web?

All the organisms shown in the food web are supported by a single tree. Thousands of primary consumers feed from the tree. The number of secondary consumers is smaller and the number of tertiary consumers is smaller still. The numbers of consumers at each level can be drawn in a **pyramid of numbers**. The pyramid of numbers for the tree looks unusual (see below).

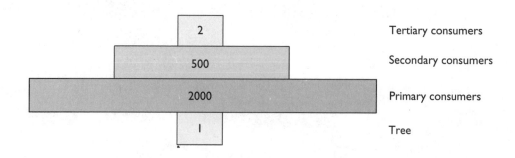

The arrows in a food chain and a food web show the flow of energy along each chain. This energy is stored in the chemicals that make up each organism. When a woodpecker eats a beetle, some of the chemicals obtained from the beetle are used for growth and repair to the woodpecker's body. These chemicals contain substances such as nitrogen which are essential for the manufacture of proteins in the organism. Sometimes chemicals that the organisms do not need are also passed up the food chain. This can happen when plants and foliage are sprayed with chemicals such as insecticides. Rain water carries the insecticide into the soil and then into streams and rivers and eventually to the sea. Industrial waste containing poisons is also discharged into the sea.

The poisons are taken up by animals and plants low down in the food chain; these plants and animals are usually unharmed by the low concentration of poisons in them. The poisons become more concentrated in the bodies of animals higher up the food chain. Big animals eat lots of smaller animals so they take in a lot of poison. The animals do not get rid of all the poison they take in when feeding, so the amount of poison in their bodies can build up to a lethal dose. The diagram shows how a food chain in a lake can be affected when insecticide pollutes the water.

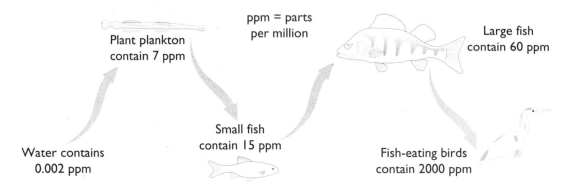

ppm = parts per million

Plant plankton contain 7 ppm

Large fish contain 60 ppm

Water contains 0.002 ppm

Small fish contain 15 ppm

Fish-eating birds contain 2000 ppm

Now test yourself

6 Forty years ago farmers started to spray their crops with a chemical called DDT to kill insect pests. Fish in the local rivers were unaffected but five years later the population of otters began to decline. Otters live near rivers and feed on fish.

(a) Describe how the DDT got into the rivers.

(b) Explain why the DDT did not affect the otters' food supply.

(c) Explain why the population of otters was not affected until five years after farmers started to use the chemical DDT.

*L*ife Processes and Living Things

At the start of Key Stage 3 the majority of pupils will have reached at least Level 4 in Science. By the end of Key Stage 3 most pupils should be within the range of Levels 4–7. Levels 5–6 are the target for 14-year-olds. Level 8 is the standard reached by very able pupils.

Use our checklist to assess the Level reached, by ticking the skills that have been mastered.

Level 4

☐ Demonstrate knowledge and understanding of aspects of life processes and living things drawn from the Key Stage 2 or Key Stage 3 programme of study.

☐ Use scientific names for some major organs of body systems, such as the circulatory system, and identify the position of these organs in the human body.

☐ Identify organs, such as petal, stamen or stigma, of different plants observed.

☐ Use keys based on observable external features to help to identify and group living things systematically.

☐ Recognize that feeding relationships exist between plants and animals in a habitat, and describe these relationships, using food chains and terms such as predator and prey.

Level 5

☐ Demonstrate an increasing knowledge and understanding of aspects of life processes and living things drawn from the Key Stage 2 or Key Stage 3 programme of study.

☐ Describe the main functions of some organs of the human body, such as the heart, and of the flowering plant, such as the petal, stamen or stigma.

☐ Explain how these functions are essential to the organism.

☐ Describe the main stages of the life cycles of humans and flowering plants and point out similarities.

☐ Recognize that there is a great variety of living things and understand the importance of classification.

☐ Explain that different organisms are found in different habitats because of differences in environmental factors, such as the availability of light or water.

Level 6

☐ Use knowledge and understanding drawn from the Key Stage 3 programme of study, to describe and explain life processes and features of living things.

☐ Use appropriate scientific terminology when describing life processes, such as respiration or photosynthesis, in animals and plants.

☐ Distinguish between related processes, such as pollination or fertilization.

☐ Describe simple cell structure and identify differences between cells, such as differences in structure between simple animal and plant cells.

☐ Describe some of the factors that cause variation between living things.

☐ Explain that the distribution and abundance of organisms in habitats are affected by environmental factors, such as the availability of light or water.

Level 7

☐ Use knowledge and understanding of life processes and living things drawn from the Key Stage 3 programme of study, to make links between life processes in animals and plants and the organ systems involved.

☐ Explain the processes of respiration and photosynthesis in terms of the main underlying chemical change.

☐ Use knowledge of cell structure to explain how cells, such as the ovum, sperm or root hair, are adapted to their functions.

☐ Identify characteristic variations between individuals, including some features, such as eye colour, that are inherited and others, such as height, that can also be affected by environmental factors.

☐ Construct models, such as food webs or pyramids of numbers, to represent feeding relationships, and explain how these relationships affect population size.

Level 8

☐ Demonstrate an extensive knowledge and understanding of life processes and living things drawn from the Key Stage 3 programme of study, in describing how biological systems function.

☐ Relate knowledge of the cellular structure of organs to the associated life processes, such as the absorption of food in the digestive system or gas exchange in the lungs.

☐ Recognize, predict and explain changes in biological systems, such as the effect of increased carbon dioxide concentration on the growth of greenhouse crops, or the consequences of smoking for organ systems.

☐ Explain how characteristics can be inherited by individuals and apply this knowledge to contexts such as selective breeding.

☐ Predict the short-term and long-term effects of environmental change on ecosystems and use understanding of such systems to justify predictions.

Exceptional performance

☐ Demonstrate both breadth and depth of knowledge and understanding of the Key Stage 3 programme of study and draw on aspects of the Key Stage 4 programme of study when describing and explaining how biological systems function.

☐ Recognize that organisms respond to change to maintain their internal environment and describe ways in which this is achieved.

☐ Relate understanding of the life processes of reproduction and growth to the processes of cell division.

☐ Use understanding of genetics to explain a variety of phenomena, such as mutation or the production of clones.

☐ Recognize the importance of quantitative data, such as that obtained from pyramids of biomass, when describing and explaining patterns of nutrient transfer within an ecosystem.

CHAPTER 7

Classifying materials

S OLIDS, LIQUIDS, AND GASES

All materials can exist in different **states of matter** depending upon temperature and pressure. The three states of matter are **solid**, **liquid** and **gas**.

There are different ways of dividing materials into groups. One way is to divide them according to their states.

Water is a very common substance. You will know that water can exist in three forms:

ice   solid
water   liquid
steam   gas

Sometimes water in a gas form is called water vapour.

When liquid water is heated, it turns to steam at 100°C. The water is said to be **boiling** and this temperature is called the **boiling point of water**.

When steam is cooled down, it turns back to water. You will have seen the water which forms on a cold window in a steamy kitchen. This change back from steam to liquid water is called **condensation**.

When water is cooled, it turns to ice at 0°C. This is called **freezing and** 0°C is called the freezing point of water. At 0°C, **melting** of ice also takes place and ice turns to liquid water.

You will know that steam (or water vapour) can turn directly into a solid. This happens inside a freezer. Solid ice forms inside the freezer when the steam in the air rapidly cools. This change is called **sublimation**.

These changes of state are summarized below.

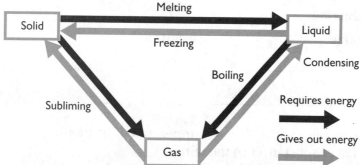

Similar changes take place with other substances. All substances can exist in three states of matter depending upon conditions. These are solid, liquid and gas.

The typical properties of solids, liquids and gases are compared in the table on the next page.

Property	Solid	Liquid	Gas
volume	definite	definite	fills the whole container
shape	definite	takes the shape of the bottom of the container	takes the shape of the whole container
density	high	medium	low
expansion on heating	low	medium	high
ease of compression	very low	low	high
movement of particles	slow	medium	fast moving particles

P REDICTING THE STATE OF A SUBSTANCE

The table below gives information about the melting and boiling points of some common substances at atmospheric pressure. Assume for this exercise that room temperature is 20°C.

Substance	Melting point °C	Boiling point °C
hydrogen	−259	−253
nitrogen	−214	−196
oxygen	−219	−183
ethanol	−117	78
ammonia	−78	−33
mercury	−39	357
bromine	−7	58
sodium	78	890
iodine	114	183
sulphur	119	445
zinc	419	908
potassium chloride	776	1427
sodium chloride	801	1420
copper	1083	2582
iron	1539	2887

A substance will be a solid at room temperature if its melting point is above 20°C. Look at the list of substances in the table.

Which **eight** substances are solid at room temperature?

A substance will be a liquid at room temperature if the melting point is below 20°C but the boiling point is above 20°C.

Which **three** substances in the table are liquid at room temperature?

A substance is a gas at room temperature if both the melting and boiling points are below 20°C.

Which **four** substances in the table are gases at room temperature?

You should now be able to work out whether any substance is a solid, liquid or a gas at a given temperature provided that you are given its boiling and freezing points.

Now test yourself

1 The chart shows the melting and boiling points of three of the substances in the table – mercury, iodine and ammonia. A horizontal dotted line has been drawn at 20°C. The horizontal line goes through the liquid portion of the mercury bar, the solid portion of the iodine bar and the gas portion of the ammonia bar. This shows that, at room temperature, mercury is a liquid, iodine is a solid and ammonia is a gas.

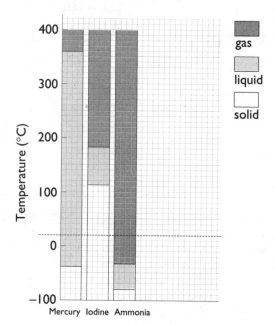

Draw a bar for ethanol and show that ethanol is a liquid. How would the states of these materials be different if they were taken to a place where the temperature was −60°C?

2 A space probe visits another planet and finds solid sulphur dioxide and solid methane on the surface of the planet. Information about these substances is found in the table.

Material	Melting point °C	Boiling point °C
sulphur dioxide	−73	−10
methane	−182	−155

What can you say about the temperature of this planet?

3 This question is about some solid crystals changing state and becoming a liquid. The crystals were heated in a test tube using a water bath until they melted. A thermometer was put into the liquid and the test tube removed from the water bath. The test tube and contents cooled. The temperature was recorded every half minute and the results are shown in the figure.

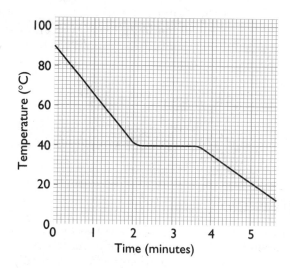

(a) Draw a diagram of the apparatus used
   (i) to heat the test tube to melt the crystals
   (ii) during the cooling of the liquid.

(b) What was the temperature after two minutes?

(c) At which temperature did the liquid turn to a solid?

C OMPOSITION OF MATTER

If you stood in the middle of the desert, you would see nothing but sand stretching in all directions. It would only be when you looked closely, and picked up the sand, that you would realize that what you see is not a solid mass but is made up of billions and billions of very tiny grains. Together the grains of sand look like a solid mass. In a similar way, all matter is made up from very tiny particles called **atoms**.

This idea is not a recent one. In about 420 BC, an ancient Greek called Democritus suggested that matter was made up of atoms. But his idea was ahead of the times and not accepted.

In 1808 John Dalton revived the idea. He proposed that matter was made up of tiny, indivisible particles called atoms, and that atoms of different materials were different from each other. For example, atoms in a block of iron were different from atoms in a block of carbon.

Atoms could be joined together in different ways to produce all the materials we know. Later in the 19th and 20th centuries, scientists realized that atoms themselves were all made up from even smaller particles called protons, electrons and neutrons.

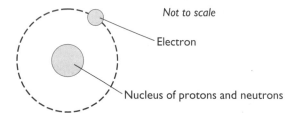

Not to scale

Electron

Nucleus of protons and neutrons

JOHN DALTON (1766–1844)

John Dalton was the son of a poor Quaker weaver who lived in Eaglesfield near Cockermouth in Cumberland. He loved the countryside and liked to study the weather. He was very bright and went to the local Quaker school. He took over there as teacher when he was only 12 years old.

When Dalton was 27, he decided that teaching was getting in the way of his scientific studies. He resigned his teaching post in Manchester and joined the Manchester Literary and Philosophical Society. Here he devoted much of the rest of his life.

From childhood, Dalton made a daily note of the weather and collected over 200 000 observations on it.

Many of Dalton's experiments were rather inaccurate and thus it is his theoretical work that will be remembered, especially his work on the atomic theory. Dalton also studied colour-blindness, probably because he was a sufferer.

He lived a simple life with regular patterns of work and play. He went on holiday to the Lake District once a year and played bowls every Thursday!

Manchester loved Dalton. At his funeral, 40 000 people filed past his coffin and the public paid for a statue which can still be seen in Manchester Town Hall. There are many other reminders in Manchester of the life of John Dalton.

(a) Scientists often study subjects because they particularly interest them. Describe three things studied by Dalton because they interested him.

(b) Dalton is remembered particularly as a **theoretical** scientist. Why do you think this is so?

ARRANGEMENT OF PARTICLES IN SOLIDS, LIQUIDS AND GASES

The figures below show a simple representation of particles in a solid, a liquid and a gas.

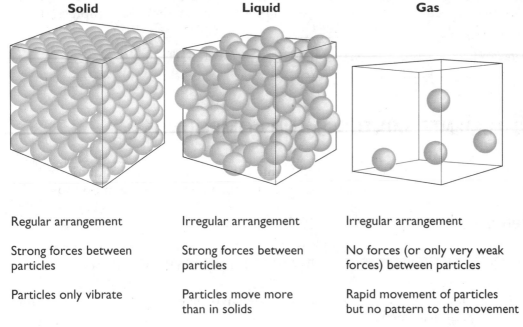

Solid	Liquid	Gas
Regular arrangement	Irregular arrangement	Irregular arrangement
Strong forces between particles	Strong forces between particles	No forces (or only very weak forces) between particles
Particles only vibrate	Particles move more than in solids	Rapid movement of particles but no pattern to the movement

These drawings show some important points.

① Particles are usually regularly arranged in solids, but are irregular in liquids and gases.
② Generally, particles are more closely packed in solids than in liquids, and more closely packed in liquids than in gases.

The diagrams, however, cannot show that the particles are moving! In a solid the particles are not moving very much: they are really only vibrating about fixed points. It is rather like being in a very crowded room and trying to get to the door! In gases the particles are moving rapidly and in all directions. The particles in a gas collide frequently with each other and with the walls of the container. In liquids there is more movement of particles than in solids but less movement than in gases. There is no pattern to the movement of particles in solids, liquids and gases. It is said to be **random movement.**

EVAPORATION

If you leave a saucer of water in a warm room for a couple of days you will find that the saucer is empty. The water has not boiled: evaporation has taken place.

Evaporation, like boiling, involves a change from liquid to gas but occurs at any temperature, not necessarily at the boiling point. To understand this, we imagine that the saucer of water is made up of millions of tiny water particles that escape into the room as evaporation occurs.

When a liquid boils, the particles are given more energy and they break away from the liquid and move faster. They move apart and occupy more space than in the liquid.

One gram of water (1 cm^3) produces over 1000 cm^3 of steam.

E FFECT OF PRESSURE ON A GAS

If you trap some air in a bicycle pump, it is easy to push in the plunger and compress the gas. The particles are forced closer together as shown in the figure. Gases decrease in volume and increase in pressure when they are compressed.

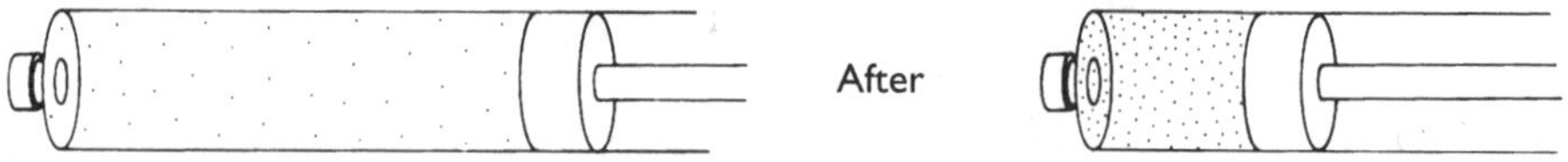

Before ████ After ████

E LEMENTS, MIXTURES AND COMPOUNDS

You know that materials can be divided into three groups – solids, liquids and gases. Materials can also be classified as elements, mixtures and compounds.

Elements

All pure substances are made up from one or more of 109 elements. These are joined together in different ways to give all of the substances in the world around us.

Hydrogen and oxygen are two elements. When hydrogen and oxygen are combined together water is formed.

An element is a pure substance which cannot be split up into anything simpler by chemical reactions. Many of these elements are found in nature but some are made in factories.

The table gives some of the common elements. For each element there is a chemical symbol which is one or two letters. Symbols are used as an abbreviation for the element.

Metallic element	Symbol	Non-metallic element	Symbol
aluminium	Al	bromine	Br
calcium	Ca	carbon	C
copper	Cu	chlorine	Cl
iron	Fe	fluorine	F
lead	Pb	helium	He
lithium	Li	hydrogen	H
magnesium	Mg	iodine	I
potassium	K	nitrogen	N
silver	Ag	oxygen	O
sodium	Na	phosphorus	P
zinc	Zn	sulphur	S

Most of the known elements are solids and metals. There are only two liquid elements at room temperature and atmospheric pressure: bromine is a liquid non-metal; and mercury is a liquid metal. The elements in the left-hand column of the table are all metals. Those in the right-hand column are non-metals. The only elements that are gases at room temperature and atmospheric pressure are hydrogen, helium, nitrogen, oxygen, fluorine, neon, chlorine, argon, krypton, xenon and radon.

All of the elements are shown in the **periodic table**.

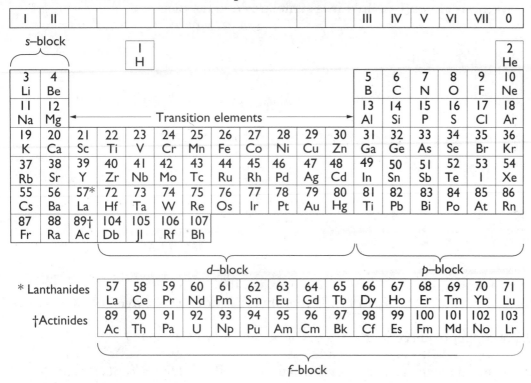

I	II												III	IV	V	VI	VII	0

s–block

		1 H		2 He

3 Li	4 Be												5 B	6 C	7 N	8 O	9 F	10 Ne
11 Na	12 Mg	←	Transition elements	→									13 Al	14 Si	15 P	16 S	17 Cl	18 Ar
19 K	20 Ca	21 Sc	22 Ti	23 V	24 Cr	25 Mn	26 Fe	27 Co	28 Ni	29 Cu	30 Zn		31 Ga	32 Ge	33 As	34 Se	35 Br	36 Kr
37 Rb	38 Sr	39 Y	40 Zr	41 Nb	42 Mo	43 Tc	44 Ru	45 Rh	46 Pd	47 Ag	48 Cd		49 In	50 Sn	51 Sb	52 Te	53 I	54 Xe
55 Cs	56 Ba	57* La	72 Hf	73 Ta	74 W	75 Re	76 Os	77 Ir	78 Pt	79 Au	80 Hg		81 Ti	82 Pb	83 Bi	84 Po	85 At	86 Rn
87 Fr	88 Ra	89† Ac	104 Db	105 JI	106 Rf	107 Bh												

d–block   *p*–block

* Lanthanides

57 La	58 Ce	59 Pr	60 Nd	61 Pm	62 Sm	63 Eu	64 Gd	65 Tb	66 Dy	67 Ho	68 Er	69 Tm	70 Yb	71 Lu

†Actinides

89 Ac	90 Th	91 Pa	92 U	93 Np	94 Pu	95 Am	96 Cm	97 Bk	98 Cf	99 Es	100 Fm	101 Md	102 No	103 Lr

f–block

All elements are made up from tiny particles called **atoms**. In a sample of an element all of the atoms contain the same number of **protons**. A sample of magnesium is made up from magnesium atoms; all magnesium atoms contain 12 protons.

Mixtures

Elements can be mixed together to form a **mixture**. For example, iron and copper powders can be mixed together to form a mixture. The mixture can be separated with a magnet.

If you look carefully at the mixture with a hand lens you will be able to see pieces of iron and copper. The mixture has all of the properties of iron and copper.

Iron

Compounds

Certain mixtures of elements react together or **combine** to form **compounds**.

For example, a mixture of hydrogen and oxygen explodes and forms droplets of water.

The formation of a compound from its constituent elements is sometimes called

synthesis. Iron(II) sulphide, the compound formed when the elements iron and sulphur combine, has entirely different properties from iron and sulphur. It is extremely difficult to get iron and sulphur back from iron(II) sulphide.

The synthesis of iron(II) sulphide is summarized in the diagram.

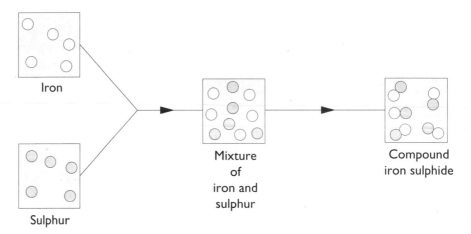

Iron

Sulphur

Mixture
of
iron and
sulphur

Compound
iron sulphide

A compound has a fixed composition by mass. If black copper oxide is formed by different methods, the mass of copper is always four times the mass of oxygen.

50 g of copper oxide contains 40 g of copper and 10 g of oxygen.

The chemical name of a compound will usually tell you the elements that are combined in the compound. If a compound ends in '-ide', the compound contains only two elements. For example:

sodium chloride   sodium and chlorine
copper(II) oxide   copper and oxygen

There are exceptions to this rule. Sodium hydroxide is composed of **three** elements: sodium, oxygen and hydrogen.

If a compound ends in '-ate' the compound contains oxygen. For example:

calcium carbonate   calcium, carbon and oxygen
copper(II) sulphate   copper, sulphur and oxygen
sodium hydrogensulphate   sodium, hydrogen, sulphur and oxygen

In a compound there may be groups of a few atoms (called **molecules**) or large arrangements (**giant structures**) of atoms or ions. The following tests help to show which is present:

1 **Heating** A compound composed of small molecules will have low melting and boiling points. A giant structure of atoms or ions will have high melting and boiling points.

2 **Conductivity of electricity** On melting, a giant structure of atoms will not conduct electricity but a giant structure of ions will conduct electricity.

Two examples of compounds with different structures are methane and silicon oxide (sand). Methane is made up of molecules each containing one carbon atom combined with four hydrogen atoms. Silicon oxide is a solid with a high melting point. It has a giant structure of atoms.

Methane molecule

Sand giant structure (silicon dioxide)

X- RAY CRYSTALLOGRAPHY

X- ray crystallography is a branch of science which requires great patience and accuracy.

A beam of X-rays is passed through a crystal. The X-rays are bent and form a pattern on a photographic plate. From this photograph, scientists can work out the arrangements of particles in the crystal.

Although X-ray crystallography owes much to the work of Lawrence Bragg, three of the most famous X-ray crystallographers have been women.

Rosalind Franklin (1920–1958) pioneered X-ray techniques of three-dimensional crystals. In 1951 she was invited to set up an X-ray unit at the biophysics department at King's College, London. One of the projects of the department was to analyse the structure of DNA, a vital chemical which passes on genetic information from one generation to another. At the same time, Francis Crick and James Watson were studying DNA. They were building up models of DNA molecules.

Unknown to Rosalind, one of her papers and her best X-ray photograph of DNA were shown to Watson in 1952 by one of her colleagues at King's, Maurice Wilkins. The photograph clearly showed that DNA had a coiled structure. This was the information that Crick and Watson needed. They were then able to make an accurate model of DNA.

The scientific magazine, *Nature*, published three papers in 1953 about the structure of DNA. One paper was by Crick and Watson, one by Wilkins and one by Rosalind Franklin.

In 1962, Crick and Watson and Wilkins were awarded the Nobel Prize for their work on DNA. In a book in 1965, Watson gave no credit to Rosalind Franklin for her part in the discovery.

Kathleen Lonsdale (1903–1971) became the first woman to be elected to the Royal Society in 1957. Her studies in X-ray crystallography included the structure of benzene and diamonds found in meteorites. Indeed, these diamonds were called Lonsdaleite.

In 1964, Dorothy Hodgkin was awarded the Nobel Prize for her work in X-ray crystallography. Her work included the structure of a protein, penicillin and vitamin B_{12}.

(a) Name three famous women X-ray crystallographers.
(b) What was Rosalind Franklin's contribution to the structure of DNA?
(c) In science, there have often been disagreements about who had done what in making a discovery. Find another example in this book.

MELTING, DISSOLVING AND DIFFUSION

Melting When a solid is melted, the regular arrangement of particles breaks down and the particles become freer and able to move.

Dissolving If a beaker is completely filled with water and salt is then slowly added to the water, the salt dissolves. The water does not overflow from the beaker, however. The salt is made up of a giant structure of ions. When the salt dissolves, the structure of ions breaks up and the ions fill spaces between the water particles.

Diffusion If you open a bottle of perfume in a room, the smell of the perfume soon spreads throughout the room. The perfume is made up from millions of tiny particles which can move around the room. Diffusion is the movement of a gas to fill any space in which it is put.

Diffusion can be demonstrated in the laboratory by putting a gas jar filled with air above a gas jar filled with heavier, red-brown bromine vapour.

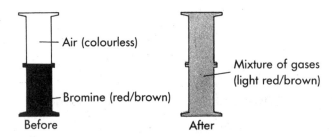

After a few minutes, the contents of the two gas jars look the same. The bromine particles have spread out evenly into both gas jars. This movement of particles is called **diffusion.**

Diffusion also takes place in liquids, though more slowly. This is because the particles in a liquid are moving more slowly than in the gas. If a purple crystal of potassium manganate(VII) (sometimes called potassium permanganate) is dropped into a beaker of water, the crystal dissolves and diffusion takes place. After some hours the whole solution is a pale pink colour. One small crystal of potassium manganate(VII) must contain enough small particles to spread out and fill all of the water.

S EPARATING MIXTURES

A pure chemical is chemical which does not contain **impurities**. A pure chemical has a definite melting point. The presence of impurities lowers the melting point and causes the substance to melt over a range of temperature.

When salt is added to water and the mixture is stirred, the salt dissolves and forms a salt **solution**. The salt disappears from view. It is still there, however, because the water tastes salty.

Now test yourself

4 The table shows the results of an experiment to compare the solubilities of common substances in water and meths.

Common substance	Solubility in water	Solubility in meths
salt	soluble	insoluble
iodine	slightly soluble	soluble
naphthalene	insoluble	slightly soluble
copper sulphate	soluble	insoluble
sulphur	insoluble	insoluble
chalk	insoluble	insoluble
charcoal	insoluble	insoluble
sugar	soluble	insoluble
sodium hydrogencarbonate	soluble	insoluble

What can you conclude from these results?

A substance which dissolves to make a solution is said to be **soluble**. A substance which does not dissolve is said to be **insoluble** (e.g. sand).

The substance which is dissolved is called the **solute** (e.g. salt in this case) and the substance which does the dissolving is called the **solvent** (e.g. water). Water is the commonest solvent but there are many others (e.g. meths, hexane).

The following rules about solubility are worth remembering:

1. Most solids dissolve better in hot water than in cold water.
2. Liquids either dissolve in water (**miscible**)or do not mix with water (**immiscible**). Changing the temperature does not affect this.
3. Gases dissolve better in cold water than in hot water.

Solubility

If you keep dissolving a solute in water at a particular temperature you will reach a stage where no more will dissolve. The resulting solution, which has the maximum amount of solute dissolved at a particular temperature, is called a **saturated solution**.

The **solubility** of a solute is the number of grams of the solute which will dissolve in 100 g of water at a particular temperature.

Most solids dissolve better in hot water than cold water. The graph shows how the solubility of various solids changes with temperature. These graphs are called **solubility curves**.

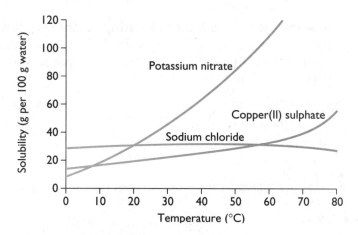

If a hot saturated solution is cooled, the solute will crystallize out.

There are a number of methods which can be used to produce pure chemicals. The method used has to be chosen carefully for each purification.

Separating a solid from a solution

An impure form of salt found underground in Cheshire is called rock salt. This consists of salt mixed with impurities, such as sand, which do not dissolve in water and are thus insoluble. The fact that the impurities do not dissolve in water is the basis of a method used to purify rock salt.

The rock salt is crushed using a pestle and mortar. The crushed rock is added to water and the mixture is stirred. The salt dissolves but the impurities sink to the bottom and form a **sediment** or **residue**. The salt solution can be removed by **decanting**.

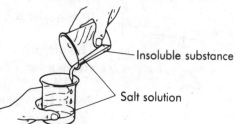

Insoluble substance

Salt solution

Alternatively, the salt solution can be removed by **filtering**. In the kitchen, flour may be sieved to remove any lumps. The flour passes through the small holes in the sieve but the lumps do not. Filtering is a very similar process. A filter paper has many very small holes through it.

The filter paper is folded into a cone shape and placed in a funnel to support it.

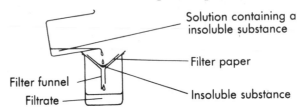

The mixture of salt solution and solid impurities is poured into the funnel. The solid impurities remain on the filter paper and the solution passes through and is collected in a beaker.

The solution collected in the beaker is called the **filtrate**.

Solid salt can be recovered from the salt solution by **evaporation**. The solution is heated in the apparatus shown until all of the water has boiled away.

Evaporating all of the water away is called **evaporating to dryness**. It does not produce good crystals, however. To get good crystals, the solution should be evaporated until a small volume of solution remains. The solution should then be left to cool, when crystals will form.

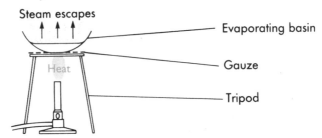

Obtaining a solvent from a solution

Evaporation is used to obtain a solute (e.g. salt) from a solution. In some cases it is important to recover the solvent from a solution. This can be done by a process called **distillation**. This is really evaporation followed by condensation.

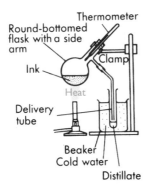

This apparatus is set up to recover some water from ink. The ink is boiled and the steam is condensed. The liquid produced is called the **distillate.** If this experiment is carried out carefully, pure water can be produced. Even so, it is difficult to stop the ink boiling over and to condense all of the steam.

The diagram shows an improved apparatus which uses a **condenser.** This piece of apparatus condenses the steam efficiently. Cold water passes through the condenser to cool down the steam.

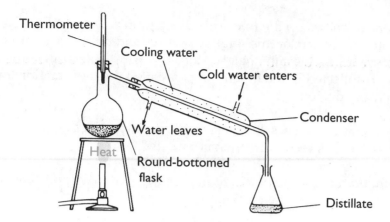

The following points should be remembered:

❶ Only steam leaves the flask. The other substances in the ink remain in the flask.

❷ The thermometer measures the temperature of the steam. The bulb of the thermometer is level with the side-arm of the flask and is not dipping in the ink. The maximum temperature recorded on the thermometer when ink is distilled should be 100°C.

❸ The condenser consists of two tubes – one inside the other. Steam passes through the inner tube and cooling water passes through the outer tube. The cooling water enters at the bottom of the condenser and leaves at the top.

❹ The condenser must slope downward so that the water which condenses runs into the receiver.

❺ The receiver should be open at the top (i.e. there should be no cork in it).

Separating mixtures of liquids

Ethanol and water mix together completely to form a single solution and are called miscible.

Hexane and water do not mix well. They form two separate layers. The top liquid is almost completely hexane while the lower liquid is almost completely water. They are said to be immiscible. The liquid in the lower layer has a greater density than the liquid in the upper layer.

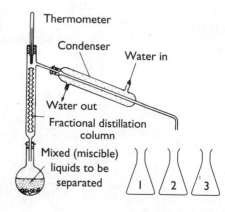

Immiscible liquids can be separated using a separating funnel. A mixture of hexane and water is placed in the funnel. After standing, the tap is opened. The water layer runs out through the tap. The hexane layer remains in the funnel and can be run out into another beaker.

Miscible liquids are much more difficult to separate but their separation is important in industry. Mixtures of miscible liquids can be separated by **fractional distillation** if the boiling points of the liquids are not too close together.

A mixture of ethanol (boiling point 78°C) and water (boiling point 100°C) can be separated by fractional distillation. The figure shows apparatus suitable for the fractional distillation of a mixture of ethanol and water in the laboratory.

The mixture to be separated is placed in the flask and small pieces of broken china are put into the flask to ensure that the liquid does not boil over.

The flask is heated slowly with receiver number 1 in place. The ethanol starts to boil first because it has a lower boiling point. The vapour passes up the fractional distillation column. Any water vapour which gets into the column at this stage condenses and drops back into the flask. The temperature on the thermometer remains below 80°C and only ethanol distils over. The liquid collected in the first receiver is called the first **fraction** and consists almost entirely of ethanol.

When the temperature reaches 80°C, receiver 2 is put in place and the temperature rises quickly to 95°C. A second fraction is collected.

When the temperature reaches 95°C, receiver 3 is put in place and soon a large volume of liquid collects in the receiver.

What do you think fractions 1, 2 and 3 consist of ?

You are right if you said fraction 1 consists of ethanol and fraction 3 consists of water. Fraction 2 will consist of a mixture of ethanol and water. If it is done carefully there will only be a small amount of fraction 2. It is difficult to separate completely a mixture of miscible liquids by fractional distillation.

Fractional distillation is used in industry in order to:

1 separate oxygen and nitrogen from air by fractional distillation of liquid air.
2 refine petroleum to produce valuable products.
3 concentrate ethanol in whisky production.

Chromatography

Chromatography is relatively simple. It is used to separate mixtures of substances dissolved in a solvent. It can also be used to identify substances. The simplest form of chromatography is called **paper chromatography**.

Paper chromatography is often used to separate mixtures of inks or dyes. It relies upon the different rates at which the dyes spread across a piece of filter paper.

A spot of the mixture of dyes is placed in the centre of a piece of filter paper. When the solvent travels up the wick it reaches the spot. The spot spreads out on the piece of filter paper. Each dye spreads out at a different rate depending upon the relative liking of the dye for the solvent and the paper. Each dye in the original mixture produces a different ring. In this example the original mixture contains two dyes.

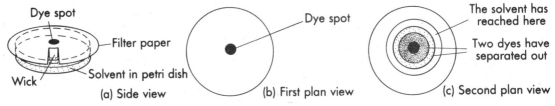

(a) Side view (b) First plan view (c) Second plan view

In practice, a square sheet of filter paper is often used. The sample spots are put on the base line and the paper is dried thoroughly. The paper is then coiled into a cylinder and put into a tank with a lid.

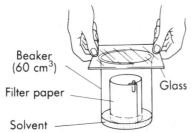

At the bottom of the tank is a small amount of solvent. The solvent travels up the filter paper and the spots are separated. When the solvent has nearly reached the top of the filter paper, the paper is removed and the position of the solvent marked.

The paper is then dried. The figure shows an example of this form of chromatography.

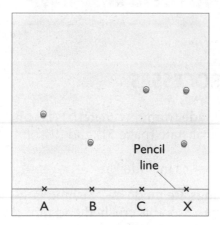

Which of the dyes A, B and C are present in X?

Chromatography can be used to identify the dyes in food to ensure they are permitted dyes. It can also be used to diagnose medical problems. For example, some people are unable to use the amino acids that they eat to build up proteins. By carrying out chromotography on their urine, the excess of amino acids leaving the body can be detected.

METALS AND NON-METALS

Elements can be divided into metals and non-metals. The table shows physical properties of a typical metal and a typical non-metal.

Metals	Non-metals
solid at room temperature	solid, liquid or gas at room temperature
shiny	dull
high density (heavy)	low density (light)
conduct heat and electricity	do not conduct electricity
can be beaten into a thin sheet (it is malleable) or drawn into a wire (it is ductile)	easily broken (brittle)
sometimes magnetic	never magnetic

Using **physical properties** to decide whether an element is a metal or non-metal can be unreliable. For example, iodine (a non-metal) is shiny like a metal and carbon (in the form of graphite) is a good conductor of electricity.

The best way of deciding whether an element is a metal or a non-metal is to use **chemical properties**. If a piece of an element is burned in oxygen an oxide is produced.

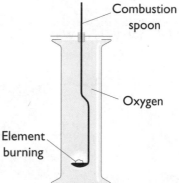

If the oxide is then tested with Universal indicator (see page 83), the pH of the oxide can be found. If the pH is less than 7 (acidic), the oxide produced is acidic and the element is a non-metal. If the pH is 7 (neutral) or greater than 7 (alkaline), the element is a metal.

CHAPTER 8

*C*hanging materials

B Y PHYSICAL PROCESSES

Many new materials are made from raw materials by physical processes. Physical processes involve changes of state (page 51) and dissolving (page 59).

Examples of physical processes include:

❶ Limestone road chippings are made by crushing large lumps of limestone.
❷ Paper is made from trees. Wood is crushed and mixed with water to form wood pulp. The wood pulp is then passed through rollers and the water is squeezed out. The result on drying is a sheet of paper.
❸ Balls of cotton which grow on the cotton plant are used to spin a continuous cotton thread.

Often materials produced by physical processes are easier to re-cycle. Usually a physical process can be reversed. For example, salt dissolved in water can be recovered by evaporation. A chemical process, such as burning a piece of wood, is not easily reversed.

Now test yourself

1 The figure summarizes two physical changes.

Dissolving sugar in water:

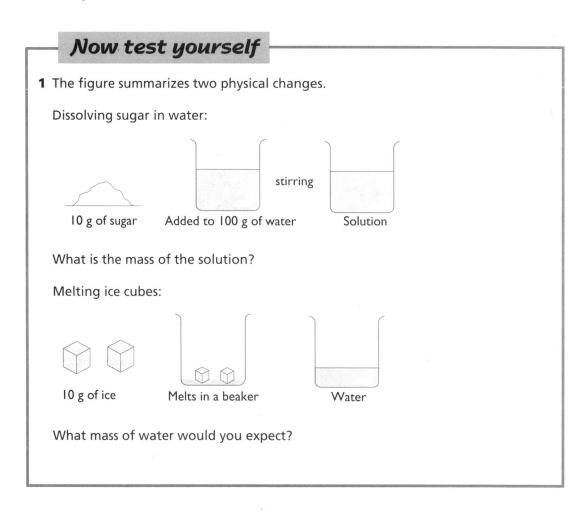

10 g of sugar   Added to 100 g of water   stirring   Solution

What is the mass of the solution?

Melting ice cubes:

10 g of ice   Melts in a beaker   Water

What mass of water would you expect?

GEOLOGICAL CHANGES

When considering the rocks of the Earth, it is difficult to understand the huge lengths of time involved. It has taken 4500 million years for the Earth to be at the stage it has reached today. You get a better idea if we use a time-chart of the history of the Earth. If we take a year to represent the life of the Earth, each day represents 12.3 million years.

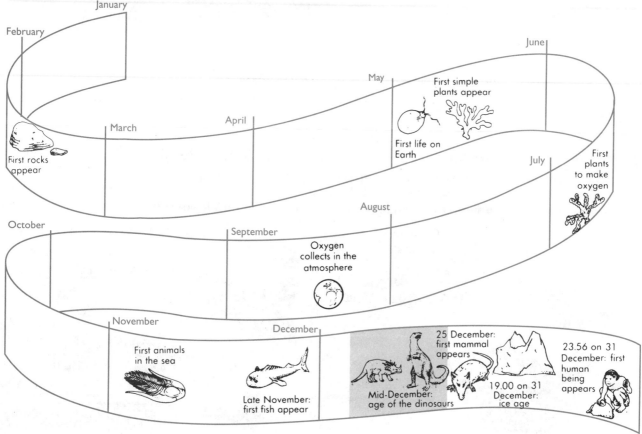

There are different types of rock in the Earth.

❶ Sedimentary rocks The rocks in the Earth are eroded or broken down in many different ways. The result is a sediment made up of many different fragments. When these layers are compressed over millions of years, sedimentary rocks are produced. Chalk and limestone are sedimentary rocks.

❷ Igneous rocks Igneous rocks are formed when molten magma from inside the Earth is cooled. The size of the crystals produced depends upon the rate of cooling: slow cooling of the liquid magma produces large crystals; rapid cooling, small crystals. Granite and basalt are different types of igneous rocks.

❸ Metamorphic rocks Very high temperatures and pressures convert sedimentary rocks into metamorphic rocks. Marble is a metamorphic rock formed from limestone. Slate is a metamorphic rock formed from mud.

Sedimentary rock – limestone

Igneous rock – granite

Metamorphic rock – slate

The rock cycle

Rocks are continually being broken down or weathered. However, the total amount of rock in the Earth's crust does not get any less. New rocks are constantly being formed in the **rock cycle**. The figure summarizes the rock cycle.

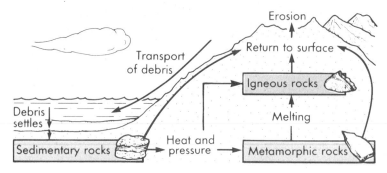

Weathering and landscaping

Modern buildings are often made of brick or concrete rather than natural stone. Where natural stone is used it is often sandstone or limestone. Both sandstone and limestone are easily weathered (or **eroded**). Limestone is badly eroded by acid rain. Erosion of stone can be seen in statues and church buildings.

Erosion is frequently caused by the effects of water. Water expands when it freezes. When water gets into a crack in a rock and then freezes, the crack is opened. Over several winters this can cause pieces of rock to fall off. A pile of pieces of rock at the bottom of a rock face is called a **scree**.

Rainfall can cause serious erosion by washing away the surface soil. This becomes more serious when forests and other surface vegetation are removed. Soil is washed off slopes and deposited on flat land. Rivers can erode the landscape by washing away rocks and soil and depositing them downstream.

The sea is a powerful eroder. It can attack cliffs and create caves, arches and other shapes.

The wind can also shape rocks. The Sphinx in Egypt has been eroded by sand carried in the wind.

Does freezing water produce enough energy to crack open rocks?

This little demonstration shows the power created by freezing water. Fill a polythene container completely with water and stand a saucer upside-down on top.

Put a heavy weight on top of the saucer. Stand everything inside a freezer and leave the water to freeze. The diagram shows what happens.

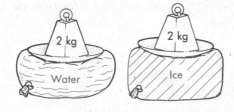

The water expands when it turns to ice and the weight is lifted up. A 4 kg mass can easily be raised, which demonstrates the tremendous forces involved. These could easily break rocks.

Soil

The weathering of rocks breaks them down into very small pieces which become part of what we call soil.

The sizes of the pieces (called grain size) varies in different soils. A sandy soil drains well because it has large grains which have big spaces between them. A clay soil soon becomes waterlogged because the spaces are small and they soon fill up with water.

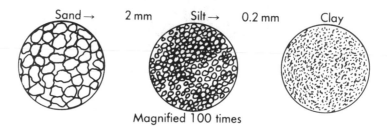

Magnified 100 times

The figure below shows a soil profile formed when soil is mixed with water and allowed to settle.

The organic horizon is dark brown. It contains dead plants and animals.

The topsoil. Rainwater washes minerals to lower levels. This process is called leaching.
Less humus found here but the minerals from 2 are deposited. Iron may precipitate out, giving the soil a yellow colour.
Small rock fragments are found here because the rock is still breaking down.
Parent bedrock.

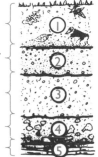

A fertile soil is good for gardening or farming. It should contain a lot of rotted vegetation called **humus** and be a **loam** soil. A loam soil contains grains of different sizes.

C HEMICAL REACTIONS

When new substances (called **products**) are made from reacting substances (called **reactants**) a **chemical reaction** takes place.

A chemical reaction is often expressed by a word equation. In the equation the reactants are on the left-hand side and the products are on the right-hand side. The arrow shows the direction of the reaction. For example:

magnesium + hydrochloric acid $\rightarrow$ magnesium chloride + hydrogen

Although the equation gives the reactants and products it gives no indication of the speed of the reaction.

In a chemical reaction the sum of the masses of the reactants equals the sum of the masses of the products, providing the masses of all the reactants are included and none of the products escape.

Oxidation and reduction

Oxidation takes place when oxygen is added or hydrogen is removed during a chemical reaction. Reduction is the opposite of oxidation (i.e. oxygen is removed or hydrogen is added). For example:

❶ **oxidation** Magnesium is oxidized (oxygen is added):

Magnesium + oxygen → magnesium oxide

❷ **reduction** Ethene is reduced (hydrogen is added):

ethene + hydrogen → ethane

Very often oxidation and reduction take place together. A reaction in which oxidation and reduction take place is called a **redox** reaction. For example, lead(II) oxide heated in a stream of dry hydrogen:

lead(II) oxide + hydrogen → lead + water

Lead(II) oxide is reduced to lead since oxygen is lost. Hydrogen is oxidized as oxygen is added. Hydrogen is called the **reducing agent** because it brings about the reduction of lead(II) oxide. Lead(II) oxide, is called the **oxidising agent** since it brings about the oxidation of hydrogen.

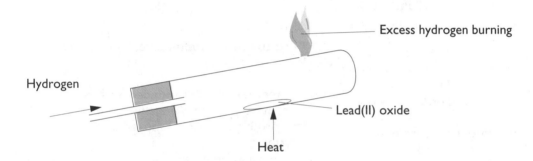

Excess hydrogen burning

Hydrogen

Lead(II) oxide

Heat

Combustion reactions

A **combustion reaction** is a reaction where a substance combines with oxygen and produces energy. When a combustion occurs **oxides** are formed.

Combustion reactions are oxidation reactions. For example, burning carbon in oxygen or burning magnesium in oxygen:

carbon + oxygen → carbon dioxide

magnesium + oxygen → magnesium oxide

When hydrocarbons burn in oxygen, different products are possible. If the hydro-carbon is burnt in excess air or oxygen, carbon dioxide and water are produced.

However, if the hydrocarbon is burnt in a limited amount of air or oxygen, carbon monoxide and water are produced.

Carbon monoxide is very poisonous because it combines with haemoglobin in the blood forming carboxy-haemoglobin. This prevents oxygen being transported around the body and leads to death. Good ventilation is necessary in a room with a gas fire to avoid carbon monoxide formation.

ANTOINE LAURENT LAVOISIER (1743–1794) AND THE THEORY OF COMBUSTION

In the 18th century, the accepted theory of combustion was called the Phlogiston Theory. According to this theory, when combustion took place the substance burning lost a substance called phlogiston. Lavoisier proved that this theory was not correct. When a substance burns it increases in mass if all the products are collected. The substance gains oxygen when it burns. The very careful work of Lavoisier disproved the Phlogiston Theory and established the new theory of combustion which we use today.

Lavoisier was born in Paris in 1743 and was given an excellent education by his wealthy parents. He studied mathematics, astronomy, chemistry and botany at Mazarin College.

He graduated in law, but was appointed Director of the Academy of Sciences in 1768. His scientific work continued here. Some of his experiments led to an improved way of making gunpowder.

He became a politician and was elected to the Assembly of Orleans in 1787. He strived to improve the living conditions for ordinary people.

However, he became a target because of his wealth when the French Revolution started. In 1794 he was arrested, tried illegally by a revolutionary court in just a few hours, and sentenced to death along with 27 others.

On the same evening, the man who did so much to improve the lives of ordinary people and advance scientific thought was guillotined in front of the masses and buried in a common grave.

Lavoisier's studies into combustion required very accurate measurements of masses and skill in handling gases and collecting them over mercury or water.

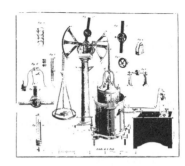

(a) The photograph shows the type of balance used by Lavoisier for weighing. What problems would this kind of balance produce?

(b) When mercury was heated in contact with air in this apparatus, oxygen was taken in by the mercury to form mercury oxide.

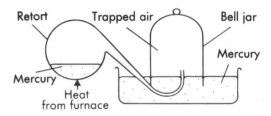

(i) Describe what happened to the mercury level in the bell jar.

(ii) The mercury oxide produced was removed from the apparatus. When the mercury oxide was produced, it decomposed and formed oxygen gas. Draw a diagram of apparatus which could be used to collect the oxygen produced.

Energy changes in chemical reactions

In a chemical reaction there is often an energy change. In a combustion reaction, or in respiration, energy is produced. A reaction which produces energy is said to be **exothermic**. For example:

sodium hydroxide + hydrochloric acid → sodium chloride + water

In photosynthesis, energy is taken in from sunlight. A reaction which takes in energy is called an **endothermic reaction**. For example, mixing together solutions of sodium carbonate and calcium nitrate:

calcium nitrate + sodium carbonate → calcium carbonate + sodium nitrate

Decomposition reactions

A substance is said to **decompose** when it splits up into two or more new substances.

Some substances decompose slowly on standing . For example, hydrogen peroxide decomposes slowly at room temperature:

hydrogen peroxide → water + oxygen

This reaction can be speeded up with a catalyst such as manganese(IV) oxide. This is an example of a **catalytic decomposition**.

When the decomposition occurs on heating, it is called **thermal decomposition**. Examples of thermal decomposition include:

❶ **heating calcium carbonate**

   calcium carbonate → calcium oxide + carbon dioxide

❷ **heating copper(II) sulphate crystals**

   copper(II) sulphate crystals → anhydrous copper(II) sulphate + water

Electrolysis reactions

Substances which cannot be decomposed by heating may be decomposed with electricity. This is called **electrolysis**.

The substance being decomposed must be molten or in solution. For example, molten lead(II) bromide can be decomposed into lead and bromine by electrolysis using the apparatus shown:

lead(II) bromide → lead + bromine

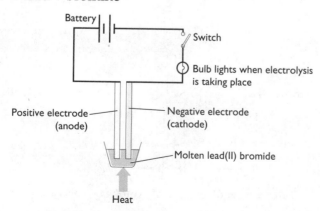

JOSEPH PRIESTLEY (1733–1804)

Joseph Priestley was a church minister with an interest in science. He was born in Fieldhead near Leeds, and became a church minister when he was 27. Three years later he began teaching classics and literature in a school in Warrington.

He wrote a history of electricity, and was admitted to The Royal Society in 1766. At this time, Priestley lived next to a brewery. He became interested in the 'heavy air' which covered the vats during fermentation. He also studied the gas given off when acid was added to chalk. He showed that the two gases were the same. This work led to his being awarded the Copley medal by The Royal Society in 1773.

Priestley discovered nitrogen in 1772. Two years later he discovered oxygen by heating mercury oxide using a large magnifying lens. The mercury oxide was decomposed into mercury and oxygen. There was some doubt about whether he had actually discovered oxygen first. In Sweden, Carl Scheele has also claimed to have discovered oxygen. Today, the credit is usually given to Priestley.

In 1780, Priestley became the minister of a church in Birmingham. His science and religion came into conflict. Although people respected his scientific writings, they disliked his religious writings so much that, in 1791, they burned down his church and house. Priestley was forced to flee to Worcester in disguise.

Later he settled at Hackney in London, but was very unhappy. He emigrated to America where three of his sons already lived. He became a minister and professor in America.

(a) Name three gases studied by Priestley.

(b) Priestley produced carbon dioxide by two different reactions. Give the two ways in which he produced carbon dioxide.

(c) Can you suggest any reasons why the work of Priestley and Scheele could go on separately without them knowing about each other?

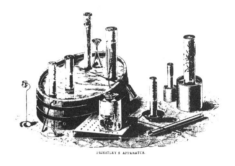

PRIESTLEY'S APPARATUS.

Lead is produced at the cathode (negative electrode) and the bromine is produced at the anode (positive electrode).

Often electrolysis of aqueous solutions is carried out. For example, the figure shows the electrolysis of copper(II) sulphate solution with carbon electrodes. During the electrolysis you would see:

❶ copper deposited as a red-brown solid on the negative electrode (cathode)
❷ bubbles of colourless gas (oxygen) produced at the positive electrode (anode)

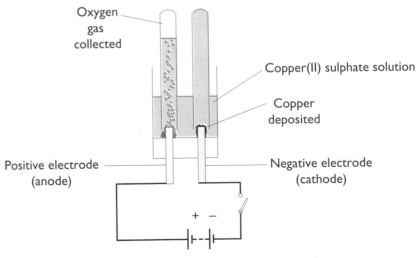

Electrolysis of copper(II) sulphate solution

❸ the blue copper(II) sulphate solution fading as the copper is deposited. The blue colour would not go if copper electrodes had been used because the copper anode would be used up to keep the blue colour.

Corrosion of metals

One of the disadvantages of some metals is their tendency to corrode. This can be an expensive problem. When a metal corrodes it reacts with oxygen and water in the air. Corrosion is an oxidation process:

metal + oxygen metal → metal oxide

There is generally a relationship between the position of a metal in the reactivity series and the way it corrodes. The higher a metal is in the reactivity series the more reactive it is, and the more quickly it will corrode.

Very reactive metals such as potassium and sodium corrode very quickly. They are stored under paraffin oil to prevent them coming into contact with air.

Corrosion of iron and steel is usually called **rusting**. It costs hundreds of millions of pounds each year in Great Britain. The diagram shows an experiment to find out what causes rusting of iron and steel to take place.

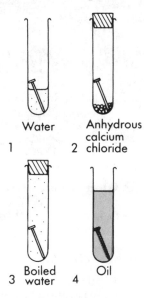

Test tube 1 An iron nail is put into water. The nail is in contact with air and water. Rusting takes place.

Test tube 2 Anhydrous calcium chloride removes all the water vapour in the air. The nail is in contact with air but not water. Rusting does not take place.

Test tube 3 The distilled water is boiled before use to remove any dissolved air. The nail is in contact with water but not air. Rusting does not take place.

Test tube 4 The nail is in oil. It is not in contact with air or water. No rusting takes place.

From these experiments, it can be concluded that air and water have to be present before rusting of iron and steel can take place. In fact, it can be shown that it is the oxygen in the air which is necessary for rusting. Other substances such as carbon dioxide, sulphur dioxide and salt speed up rusting.

Rusting of iron and steel can be reduced by:

1. **Oiling or greasing** e.g. keeping the lawnmower blades oiled over the winter
2. **Painting** e.g. iron railings
3. **Coating with plastic** e.g. washing-up racks
4. **Coating with zinc (galvanising)** e.g. metal dustbins
5. **Sacrificial protection** If a reactive metal such as magnesium is kept in contact with the iron, the magnesium corrodes instead of the iron. Although magnesium is expensive, this is a good method for preventing rust on the hull of a ship.

Factors affecting the rate of a reaction

A reaction which takes place quickly is called a fast reaction and is finished in a short time. There are a number of ways of speeding up a chemical reaction.

1. **Increasing the surface area of the solid.** Small lumps of a chemical have a much larger surface area than a single lump of the same chemical of equal mass. Powders have a very large surface area. Flour dust in a flour mill has to be carefully controlled because mixtures of flour dust and air can explode.
2. **Increasing the concentration of reacting substances.** Doubling the concentration of one of the reacting substances will often double the rate of reaction (i.e. halve the time taken for the reaction). In reactions involving gases, the concentration can be increased by increasing the pressure.
3. **Increasing the temperature.** The rate of reaction increases considerably when the temperature is increased. A temperature rise of 10°C often doubles the rate of reaction. Lowering the temperature will reduce the rate of reaction. For example, the rate of souring of milk or spoiling of food is reduced by cooling. A refrigerator or deep freezer cools the food down so that the chemical reactions which lead to spoiling are slowed down.

❹ Using a catalyst. A catalyst is a substance which alters the rate of a chemical reaction without being used up. Usually a catalyst is used to speed up reactions. For example, in the contact process to produce sulphuric acid, the catalyst is vanadium(V) oxide:

sulphur dioxide + oxygen → sulphur trioxide

Sometimes a catalyst is used to slow down a reaction. For example, additives are added to food to prevent it going bad.

Many chemical reactions taking place in living things are controlled by biological catalysts called **enzymes**. Enzymes are proteins. They have specific properties:

❶ A particular enzyme will only catalyse particular reactions, not all reactions.
❷ They only work over a limited range of temperature (e.g. enzymes which operate in the human body will work at temperatures around the normal body temperature of 37°C).
❸ When enzymes are heated they are chemically changed and then stop working.

Examples of enzymes are:

❶ amylase in saliva which breaks down large starch molecules into smaller glucose molecules
❷ enzymes in biological washing powders which remove stains in cool water
❸ enzymes in yeast which convert sugar into ethanol during fermentation.

Fermentation

Fermentation is an important industrial process which converts sugar or starch solution into ethanol and carbon dioxide gas. This takes place when enzymes in yeast act on the solution at room temperature. Fermentation continues until about 10 per cent ethanol is present. The enzymes are then poisoned by the ethanol.

Fermentation is used to prepare wine from grape juice or beer from hops.

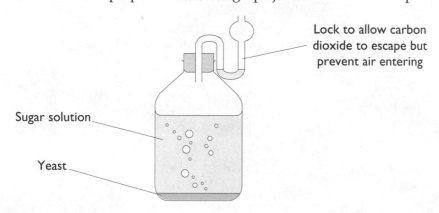

Lock to allow carbon dioxide to escape but prevent air entering

Sugar solution

Yeast

Fractional distillation of a solution of ethanol in water produces a **spirit**.

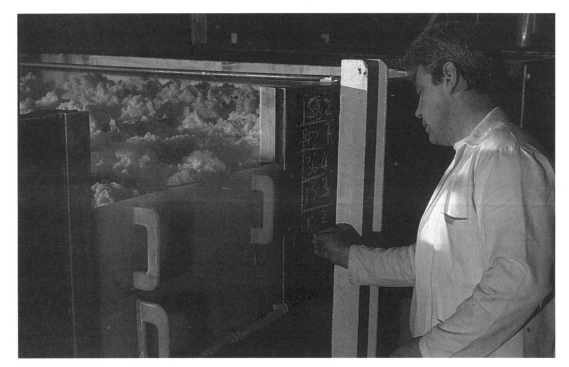

The fermentation process: this man checks the specific gravity of the liquor to monitor the speed of fermentation

T HE EARTH'S RESOURCES

The raw materials used in industry come from the Earth. Air, water, rocks, living things and fossil fuels are all raw materials. Sometimes these are called **resources**.

Now test yourself

2 Many of the materials we use at home have been made or **manufactured** using resources from the Earth. Look at some of the materials used in building a house. Which materials are natural and which manufactured?

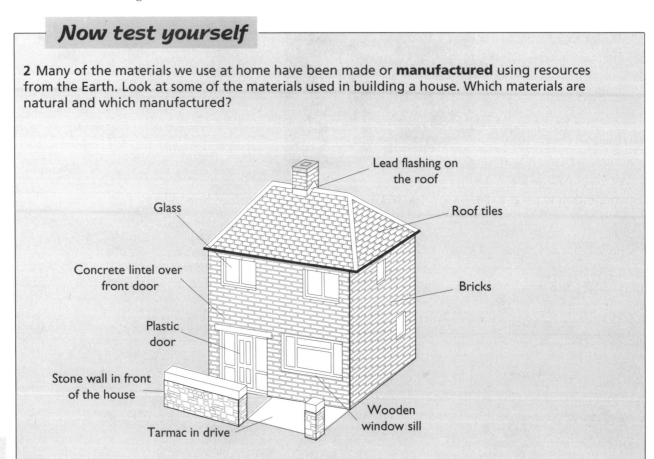

P OLYMERS

Many of the items which used to be made of metals are now made of plastic materials called **polymers**. Polymers are usually manufactured from petroleum.

For example, a modern car contains many components made from polymers.

CHASSIS	ENGINE	
brake hose	belt drive	air filter
shock absorber fluid	fan	air filter housing
bellows	timing chain drive and cover	gaskets
gaiters	fuel pipes	switch gear
	fuel pump diaphragm	cable ducting
	radiator hose	carburettor float
	clutch lining	cylinder head gasket
	fuel tank	

The parts listed are parts of the chassis and engine made from polymers. Can you list twelve other parts of the car made of polymers?

Polymers are made up from very long chain molecules. These long chains are made up by joining together many small molecules called **monomers**. There can be between 1000 and 50 000 monomer molecules linked together in a polymer chain. The diagram below summarizes the process taking place during polymerization.

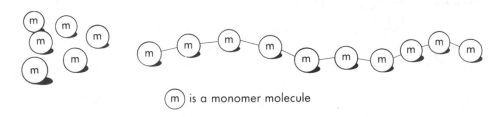

(m) is a monomer molecule

CHAPTER 9

P atterns of behaviour

R EACTIONS OF METALS

Metals react with air, water and acids. Different metals react in different ways.

Reactions of metals with oxygen

Most metals react with oxygen to form a metal oxide. Reactive metals such as potassium and sodium react without heating. Other metals such as magnesium and zinc react on heating. Copper forms only a black surface coating of copper oxide when heated in oxygen or air. Very unreactive metals such as gold do not react with oxygen.

The metal oxide formed when a metal burns in oxygen is called a **base**.

Reactions of metals with water

Potassium and sodium react with cold water to produce hydrogen gas. If a small piece of sodium is put onto water it floats and reacts quickly. An alkali (potassium hydroxide or sodium hydroxide) is produced in the water. With potassium, the hydrogen catches light.

Calcium reacts slowly with cold water to produce hydrogen and calcium hydroxide. Unlike sodium and potassium, calcium sinks in the water.

Magnesium, aluminium and zinc do not react with cold water but do react with steam. For example:

magnesium + steam → magnesium oxide + hydrogen

Iron only reacts partially with steam. Lead, copper, silver and gold do not react with water or steam.

Reactions of metals with acids

Most metals react with acids to produce a **salt** and hydrogen gas. Very reactive metals such as potassium are too reactive to try to react with acids. Very unreactive metals do not react. Examples of reactions include:

magnesium + hydrochloric acid → magnesium chloride + hydrogen

zinc + sulphuric acid → zinc sulphate + hydrogen

magnesium + nitric acid → magnesium nitrate + hydrogen

(Hydrogen is rarely produced with nitric acid due to complicated further reaction.)

The reactivity series of metals

The reactions of metals with air, water and acids can be used to arrange metals in order of decreasing reactivity. This list may be called the **reactivity series**.

The reactivity series for common metals is:

potassium

sodium

calcium

magnesium

aluminium

zinc

iron

lead

copper

silver

gold

The reactivity series is useful for predicting likely chemical reactions. For example, if iron filings are added to blue copper(II) sulphate solution, a brown solid is formed and the solution turns colourless. The iron and copper change places and free copper is formed.

copper(II) sulphate + iron → copper + iron(II) sulphate

This is called a **displacement reaction**. It takes place because iron is more reactive than copper. Iron is higher in the reactivity series than copper.

If silver powder is added to copper(II) sulphate solution there is no reaction because silver is less reactive than copper.

The Thermit reaction is a practical application of a displacement reaction. It is used to weld together long lengths of railway track from shorter lengths. A mixture of aluminium powder and iron(III) oxide powder is placed between the two ends of railway track. The mixture is set alight and a reaction takes place.

iron(III) oxide + aluminium → iron + aluminium oxide

The heat generated is enough to melt the iron which flows into the gap between the rails and welds them together.

A CIDS AND ALKALIS

Many of the substances around us are acids or alkalis. The sharp taste we get when we bite into an apple is an acid. Acids always have a sour taste, although we would be unwise to taste most of them! Acids are present in lemons, oranges, limes (i.e. citrus fruits contain citric acid). The sourness in sweet and sour chicken comes from vinegar. This contains ethanoic (acetic) acid.

There are three common mineral acids:

sulphuric acid $\qquad$ H_2SO_4

nitric acid $\qquad$ HNO_3

hydrochloric acid $\qquad$ HCl

An alkali is a base which is soluble in water. When calcium oxide (a base) is dissolved in water an alkali, calcium hydroxide, is formed:

calcium oxide + water $\rightarrow$ calcium hydroxide

Washing powders, caustic soda and ammonia solution are examples of alkalis. An alkali solution is usually soapy.

A **neutral** substance is neither acid nor alkaline (e.g. pure water, ethanol or petrol).

The simplest way of testing for an acid or an alkali is to use litmus paper or litmus solution. Litmus is a purple-coloured extract of a lichen which changes colour depending upon whether acid or alkali is added. If litmus is added to an acid, the litmus turns red. If it is added to an alkali it turns blue.

This can be summarized as: $\qquad$ Alkali $\qquad$ Acid

$\qquad$ Blue $\qquad$ Red

Litmus gives you no idea about the strength of an acid or alkali. Both vinegar and sulphuric acid turn litmus red.

Now test yourself

1 Three test tubes contain different liquids. A piece of red litmus paper and a piece of blue litmus paper are added to each test tube. The results are summarized in the table.

Liquid	Red litmus	Blue litmus
A	stays red	turns red
B	turns blue	stays blue
C	stays red	stays blue

What can you conclude about each liquid from these tests?

Universal indicator is a better test for acids and alkalis. Universal indicator is a mixture of simple indicators, and changes through several colours. From the colour you can get the pH. The colours are shown in the table.

pH	colour	acidity/alkalinity
1		
2		
3	red	
4		acid
5	orange	
6	yellow	
7	green	neutral
8	blue	
9	indigo (blue/violet)	
10		
11		alkali
12	purple	
13		

pH is a number on a scale which shows how acid or how alkaline a substance is. A substance which is neither acid nor alkaline is said to be neutral. If a solution is slightly acidic it turns Universal indicator yellow and has a pH of 6.

It is also possible to measure the pH of a solution using a pH meter as shown above.

Now test yourself

2 Suppose you wanted to measure the pH of:
   (a) blackcurrant cordial

   (b) bleach.

Why would this be difficult using Universal indicator?

Neutralization reactions

All acids contain hydrogen. This can be replaced by a metal or an ammonium ion. The substance formed when the hydrogen in an acid is replaced is called a **salt**.

When an alkali is added to an acid, the acidity is slowly destroyed. If equal amounts of acid and alkali are mixed together a neutral solution is formed. This process is called **neutralization.**

Sodium chloride (commonly called salt) is produced when sodium hydroxide (alkali) neutralizes hydrochloric acid:

Sodium hydroxide + hydrochloric acid → sodium chloride + water

There are many everyday examples of neutralization:

❶ Everyone has several hundred cubic centimetres of hydrochloric acid in the gastric juices of the stomach. This is used in the digestion of food. Minor problems of indigestion are caused by excess acid in the stomach. This can be corrected by taking antacids such as milk of magnesia (a suspension of magnesium hydroxide) or bicarbonate of soda (sodium hydrogencarbonate).

2. Lime mortar consists of a mixture of calcium hydroxide and water. This hardens when it absorbs carbon dioxide from the air. The calcium hydroxide is neutralized by acid gases in the air.

Calcium carbonate is formed:

calcium hydroxide + carbon dioxide → calcium carbonate + water

3. Many inland lakes in Scotland and Scandinavia are slowly becoming acidic because of air pollution and acid rain. Fish are dying and lakes are becoming lifeless. In an attempt to correct this, the land around the lakes is being treated with lime. As the lime is washed into the lakes it neutralizes some of the acidity.

4. Farmers have to control the pH of their soil. If the soil becomes too acidic, a good yield of crops cannot be obtained. Rain and artificial fertilizers tend to make the soil more acidic. The farmer can neutralize land by treating it with lime.

MAKING SALTS

Salts are produced when hydrogen atoms in an acid are replaced by a metal. For example:

Acid		Example of a salt	
H_2SO_4	sulphuric acid	$NaHSO_4$	sodium hydrogensulphate
		Na_2SO_4	sodium sulphate
HNO_3	nitric acid	$NaNO_3$	sodium nitrate
HCl	hydrochloric acid	$NaCl$	sodium chloride
H_2CO_3	carbonic acid	$NaHCO_3$	sodium hydrogencarbonate
		Na_2CO_3	sodium carbonate

When making a salt it is important to use the correct acid.

Salt produced	Acid used
chloride	hydrochloric acid
sulphate	sulphuric acid
nitrate	nitric acid

There are four reactions which can be used to prepare salts:

metal + acid → salt + hydrogen

metal oxide (base) + acid → salt + water

metal hydroxide + acid → salt + water

metal carbonate + acid → salt + water + carbon dioxide

The figure summarizes the stages which can be used to prepare copper sulphate from solid copper oxide. Copper sulphate is a soluble salt which crystallizes to form blue copper sulphate crystals. Each molecule of copper sulphate contains five molecules of water of crystallization.

copper oxide + sulphuric acid → copper sulphate + water

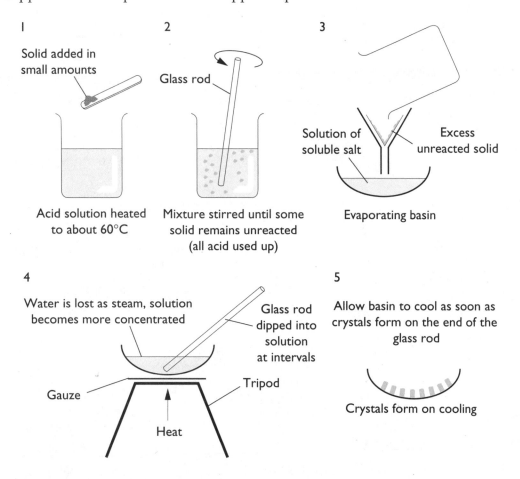

Preparation of copper sulphate from copper oxide

Now test yourself

3 Complete the following word equations for reactions producing salts:
 (a) zinc + hydrochloric acid → _____ + _____

 (b) sodium hydroxide + _____ → sodium chloride + _____

 (c) _____ + hydrochloric acid → calcium chloride + _____ + _____

 (d) _____ + _____ → magnesium sulphate + _____

E FFECT OF ACIDS IN THE ATMOSPHERE ON METALS AND ROCKS

Sulphuric acid and nitric acid can exist in the atmosphere. Sulphuric acid comes from the combustion of fuels which contain high sulphur content, e.g. coal. Nitric acid comes naturally from the action of lightning on nitrogen and oxygen and from reactions between nitrogen and oxygen in car engines.

Acids in the atmosphere speed up the corrosion of metals. Many rocks, e.g. limestone and marble, are carbonates. Acids in the atmosphere will react with carbonate rocks.

*M*aterials and their Properties

At the start of Key Stage 3 the majority of pupils will have reached at least Level 4 in Science. By the end of Key Stage 3 most pupils should be within the range of Levels 4–7. Levels 5–6 are the target for 14-year-olds. Level 8 is the standard reached by very able pupils.

Use our checklist to assess the Level reached, by ticking the skills that have been mastered.

Level 4

☐ Demonstrate knowledge and understanding of aspects of materials and their properties drawn from the Key Stage 2 or Key Stage 3 programme of study.

☐ Describe differences between the properties of different materials and explain how these differences are used to classify substances as solids, liquids and gases.

☐ Describe some methods, such as filtration, that are used to separate simple mixtures.

☐ Use scientific terms, such as evaporation or condensation, to describe changes.

☐ Use knowledge about some reversible and irreversible changes to make simple predictions about whether other changes are reversible or not.

Level 5

☐ Demonstrate an increasing knowledge and understanding of aspects of materials and their properties drawn from the Key Stage 2 or Key Stage 3 programme of study.

☐ Describe some metallic properties, such as good electrical conductivity, and use these properties to distinguish metals from other solids.

☐ Identify a range of contexts in which changes, such as evaporation or condensation, take place.

☐ Use knowledge about how a specific mixture, such as salt and water, or sand and water, can be separated to suggest ways in which other similar mixtures might be separated.

Level 6

☐ Use knowledge and understanding of the nature and behaviour of materials drawn from the Key Stage 3 programme of study, to describe chemical and physical changes and how new materials can be made.

☐ Recognize that matter is made up of particles, and describe differences between the arrangement and movement of particles in solids, liquids and gases.

☐ Identify and describe similarities between some chemical reactions, such as the reactions of acids with metals or the reactions of a variety of substances with oxygen.

☐ Use word equations to summarize simple reactions.

☐ Relate changes of state to energy transfers, in contexts such as the formation of igneous rocks.

Level 7

☐ Use knowledge and understanding drawn from the Key Stage 3 programme of study, to make links between the nature and behaviour of materials and the particles of which they are composed.

☐ Use the particle model of matter in explanations of phenomena such as changes of state.

☐ Explain differences between elements, compounds and mixtures in terms of their constituent particles.

☐ Recognize that elements and compounds can be represented by symbols and formulae.

☐ Apply knowledge of physical and chemical processes to explain the behaviour of materials in a variety of contexts, such as the way in which natural limestone is changed through the action of rainwater, or ways in which rocks are weathered.

☐ Use patterns of reactivity, such as those associated with a reactivity series of metals, to make predictions about other chemical reactions.

Level 8

☐ Demonstrate an extensive knowledge and understanding of the Key Stage 3 programme of study in describing and explaining changes to materials.

☐ Use the particle model in a wide range of contexts.

☐ Describe what happens in a range of chemical reactions and classify some, such as oxidation or neutralization.

☐ Represent common compounds by chemical formulae and use these formulae to form balanced symbol equations for reactions, such as those of acids with metals, carbonates or oxides.

☐ Apply knowledge of patterns in chemical reactions to suggest how substances, such as salts, could be made.

Exceptional performance

☐ Demonstrate both breadth and depth of knowledge and understanding of the Key Stage 3 programme of study and draw on aspects of the Key Stage 4 programme of study when describing and explaining the nature and behaviour of materials.

☐ Recognize that there are differences in the structure and bonding of different materials and relate these differences to the properties of the materials.

☐ Recognize and classify a range of chemical reactions, such as reduction or thermal decomposition.

☐ Routinely use balanced symbol equations for the reactions described.

☐ Recognize that the way in which an element reacts is related to its electronic structure.

CHAPTER 10

*E*lectricity and magnetism

E LECTROSTATICS

When a balloon is rubbed with a duster it becomes electrically charged. A charged balloon can stick to a ceiling, pick up small pieces of paper and even make someone's hair stand up!

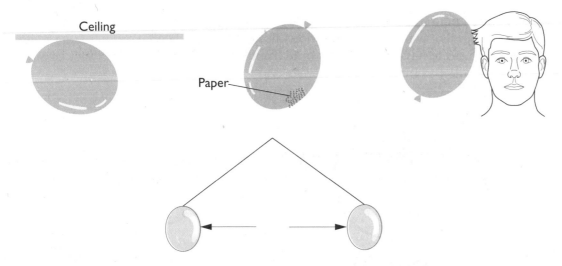

If you charge up two balloons by rubbing them with a duster, they push away from each other because they have the same kind of charge. The duster and a balloon attract each other, so they must have opposite kinds of charge.

When an insulating material such as rubber or polythene is given an electric charge, the charge stays on it because charges cannot move through insulators.

The friction between a duster and the rubber of a balloon causes electrons to be rubbed off the duster and onto the balloon. Electrons have a negative charge, so the balloon, which has been given extra electrons, has gained negative charge.

The duster has lost electrons so it now has an overall positive charge because it has more positively-charged particles than negatively-charged ones.

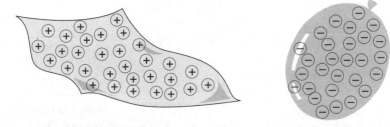

Now test yourself

1 When you walk on a nylon carpet the carpet can become negatively charged.

(a) Use the idea of the transfer of electrons to explain how this happens.

(b) What type of charge are you likely to get? Explain why.

C URRENT IN CIRCUITS

Electric currents are useful to us because they can heat things up, make things give out light and make things move. A toaster, a lamp and a vacuum cleaner all depend on electric current. Mains electricity can be dangerous and could give you a fatal shock, so you must always take care when using appliances that work from the mains.

It is safe to experiment with electricity using dry cells, but you must not try out these experiments using a car battery or mains electricity.

Using a lamp and a dry cell or a low voltage power pack, you can easily see that electricity is not like water or gas that flow along a pipe and come out when a tap is opened. You need to have a complete route from the cell to the lamp and back again. This route is called a circuit.

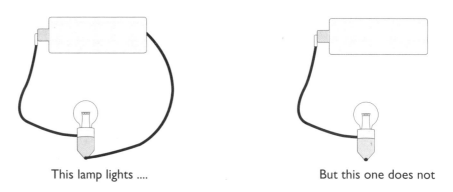

| This lamp lights .... | But this one does not |

If there is a break in the circuit, the lamp goes out. You can use a circuit like this to test which materials are conductors and which are insulators.

Air is normally an insulator. A high voltage is needed to make air conduct electricity, so an air gap in a circuit can be used as a switch. If the gap is closed using a conducting material, the circuit is switched on. The diagram below uses circuit symbols to show a lamp that can be switched on and off.

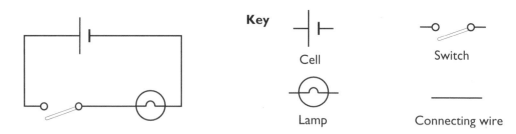

Key

Cell

Lamp

Switch

Connecting wire

Here are some more symbols that you may find useful.

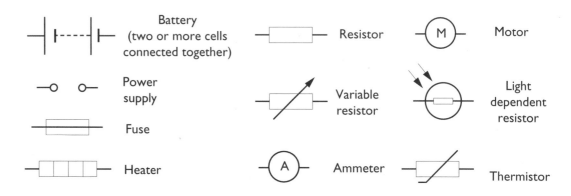

Battery (two or more cells connected together)

Power supply

Fuse

Heater

Resistor

Variable resistor

Ammeter

Motor

Light dependent resistor

Thermistor

There are two different ways of connecting two lamps to one cell or power supply.

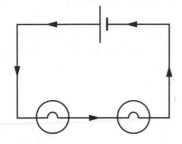 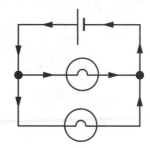

The circuit on the left is called a **series** circuit; the current from the cell passes through each lamp in turn, one after the other. In the **parallel** circuit shown on the right, the current splits at the junction before the lamps and rejoins at the junction after the lamps.

Electronic systems such as those in computers, radios and televisions contain many series circuits, but most household mains appliances are in a parallel circuit. The exception is some types of Christmas tree lights.

One problem with series circuits is that a break anywhere in a circuit turns the whole circuit off. With a parallel circuit, switches can be put in the branches of the circuit so that each switch controls just one device.

Now test yourself

2 The circuits show two designs for lighting your bedroom and your sister's bedroom.

Explain why the circuit on the right is preferable to that on the left.

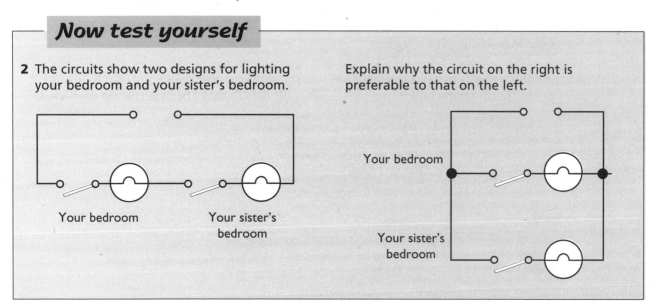

CONTROLLING CURRENT

The brightness of a lamp is an indication of the size of the electric current passing in a circuit. The current can be made bigger or smaller if a **variable resistor** is included in the circuit.

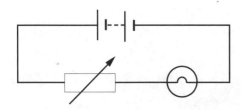 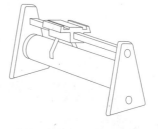

The diagram shows a common type of variable resistor and the circuit diagram shows how it is connected in a lamp-dimming circuit.

Moving the slider to one end of the variable resistor makes the lamp light at its brightest. The lamp is dimmest when the slider is moved to the opposite end.

Variable resistors can control other things besides lamps. They act as the volume control on radios and they can also be used to control the speed of an electric motor.

Now test yourself

3 Design a circuit that uses a lamp, a motor, a switch and a variable resistor. The switch should control the lamp only and the variable resistor should control the motor only.

4 Design a circuit that uses a lamp, a motor and a variable resistor. The variable resistor should control the brightness of the lamp and the speed of the motor.

MEASURING CURRENT

Sometimes the current passing in a circuit may be so small that it is not enough to light a lamp. Or it may be too large and would cause the lamp to 'blow'.

An ammeter not only detects a greater range of current than a lamp does, it also gives you more precise measurements and enables you to make comparisons. Ammeters are either digital, which are easy to read, or analogue, which involve a needle moving over a scale.

Analogue meters call for more care when taking readings, as you have to interpret the scale divisions and make judgements about readings between scale divisions.

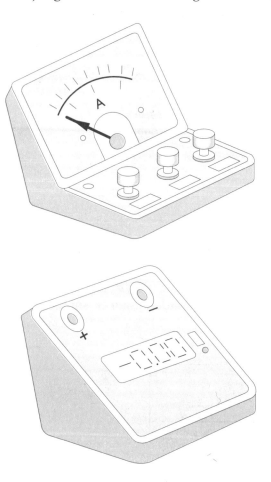

The diagrams below show how to connect an ammeter to measure the current passing into and out of a lamp.

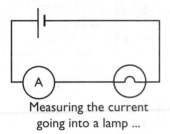

Measuring the current going into a lamp ...

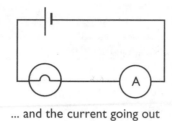

... and the current going out

Most people are surprised when they do this experiment and find that **the current coming out of the lamp is the same as that going in**. The lamp does not use up any current at all. The results obtained from the next circuit confirm this.

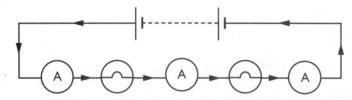

All the ammeter readings are the same; the lamps have not used any current. This may seem strange, because without a current the lamps do not work, so how do lamps produce light?

What comes out of a lamp is **energy**, in the form of heat and light. The current in a circuit transfers energy from the source, the battery or power supply, to the lamp and other components such as motors or heaters.

The energy is transferred around a circuit by moving charged particles. In metals the moving particles are electrons; these carry a negative charge. Electrons cannot move in an insulator but they are free to move around in a conductor.

The electron movement is from negative to positive, even though we always mark current directions as being from positive to negative.

When a current passes in ionized gases and in liquids, both positively and negatively charged particles are moving, but they move in opposite directions. Positive ions move to the negative electrode and negative ions move to the positive electrode.

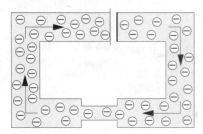

The diagram shows a model of charge flow in a circuit containing a cell and a lamp; the narrow part of the circuit represents the lamp filament. The negatively charged electrons move around the circuit from negative to positive.

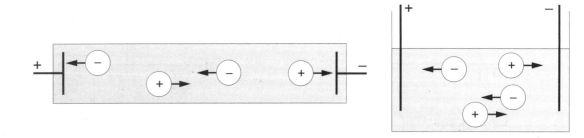

In ionized gases, such as in a fluorescent tube, and in liquids the current is carried by both positive and negative charges.

C HANGING THE CURRENT

On page 91 you saw how a variable resistor can be used to vary the current in a circuit. The size of the current that passes in a circuit depends on two things: the **voltage** of the current source and the **resistance** of the circuit.

A variable resistor works by changing the resistance; the more resistance there is, the smaller the current that passes.

When you increase the voltage that drives the current in a circuit, a bigger force acts on the moving charges. The charge travels round the circuit at a greater rate, so increasing the current.

The figure below shows that when you add more lamps in a series circuit, the current becomes less because you are increasing the circuit resistance.

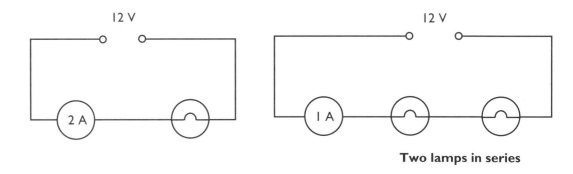

Two lamps in series

Adding more lamps in parallel causes more current in the circuit because there are more routes available for the current to pass through.

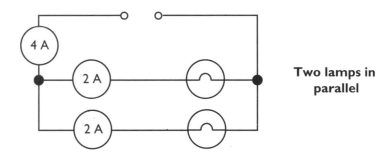

Two lamps in parallel

This circuit shows that the current splits at a junction; the sum of the currents along the branches is equal to the current from the power supply.

Now test yourself

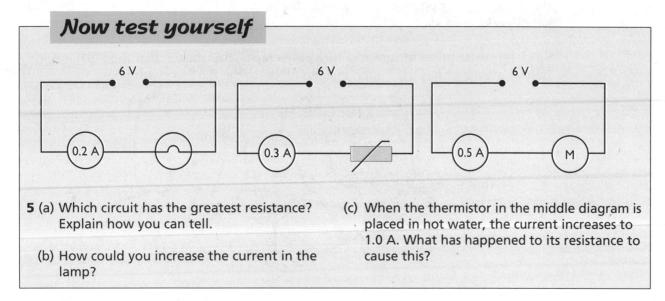

5 (a) Which circuit has the greatest resistance? Explain how you can tell.

(b) How could you increase the current in the lamp?

(c) When the thermistor in the middle diagram is placed in hot water, the current increases to 1.0 A. What has happened to its resistance to cause this?

MAGNETISM

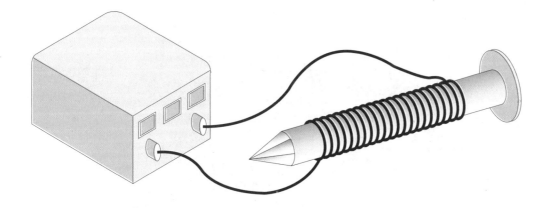

Magnets attract iron and steel and nickel but they do not affect copper, brass, chrome, magnesium or zinc. This property makes them useful for sorting ferrous (iron-based) metals from non-ferrous ones.

Iron is a 'soft' magnetic material; it is easy to magnetize but quickly loses its magnetism.

Steel is harder to magnetize but it keeps its magnetism for a long time. If you test 'copper' coins with a magnet you will find that those minted since 1992 are strongly magnetic because they are made from steel which has been copper plated.

Magnets can attract and repel other magnets. The strongest parts of a magnet are called the poles. Bar magnets have poles at the ends but slab magnets have them along the sides.

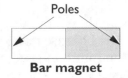

Bar magnet

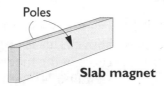

Slab magnet

Magnets have been used for centuries for navigation. A magnet that is free to turn round always points with one pole pointing towards magnetic north. This is called the north-seeking pole or north pole of the magnet. The opposite pole is called the south-seeking or south pole.

Since **opposite poles attract** and **like poles repel**, this means that the north-seeking pole of a magnet has the opposite magnetism to the Earth's north pole.

Compasses use small magnets pivoted on needles to enable them to turn round so that the poles point towards magnetic north and south. However, if a compass is placed near a magnetic material such as iron, it will point towards that instead.

When used for navigating, compasses should always be kept away from anything made from iron, nickel or steel.

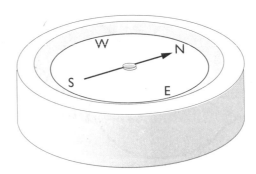

A compass is a useful tool for investigating the attractive and repulsive forces around a magnet. By looking at the direction of the force on the north-seeking pole of the compass, a set of straight lines can be drawn showing the force pattern all around the magnet.

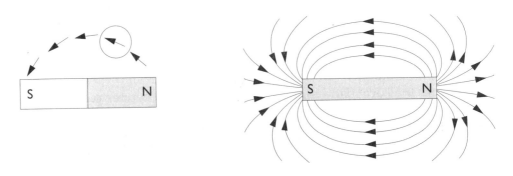

When these lines are joined up they make a pattern like that shown in the right-hand diagram above. This is called a **magnetic field pattern**, because it shows the pattern of the forces in the magnetic field. This is the region round the magnet where it pushes and pulls magnetic objects.

Although the lines are curved, it is important to realize that the force at any point is in a straight line. Magnetic fields are always drawn to show the force on the north-seeking pole of another magnet.

Iron filings act like small compasses. If you sprinkle iron filings around a magnet, they line up to show the shape of the magnetic field pattern. Where the filings line up to show similar poles, there is a repulsive field. Where there are two opposite poles, there is an attractive field. Iron filings give a quick way of investigating a magnetic field, but you need to use a compass to see the direction of the forces.

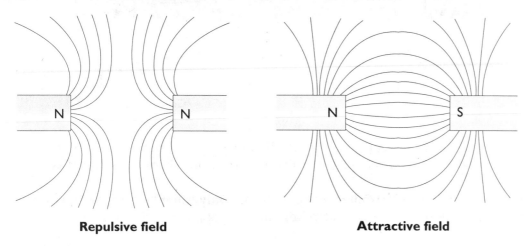

Repulsive field   **Attractive field**

Now test yourself

6 A pupil investigates the magnetic field pattern around the end of a steel bar. The diagram shows the results.

(a) What does this tell you about the steel bar?

(b) What would you expect the field pattern to look like around the opposite end of the bar? Explain why.

E LECTROMAGNETISM

Electromagnets are all around us. They are used in motors to turn the washing machine drum and the lawn mower blades. Electromagnets are also used to make a loudspeaker cone vibrate and to 'write' information in magnetic form onto audio and video tapes and computer disks.

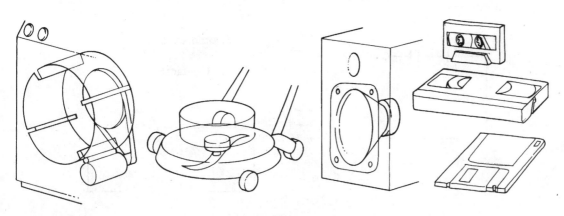

All electric currents have their own magnetic field. This means there are magnetic forces around them. We do not normally notice the effects of these forces because they are very weak. Careful observation using iron filings and a compass reveals the field patterns shown below for a straight wire and for a coil.

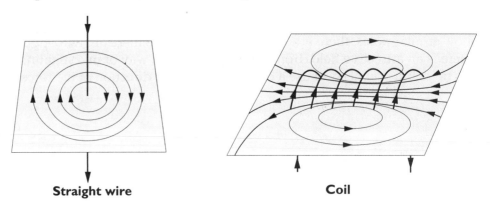

Straight wire   **Coil**

Electromagnetic forces can be made much bigger by wrapping a coil of wire around an iron core. When a current passes in the wire, the iron becomes strongly magnetized.

This makes a useful electromagnet because, not only can it be switched on and off, its strength can be varied by controlling the current passing through the coil.

The diagram shows how you can make an electromagnet that is capable of picking up magnetic objects such as paper clips.

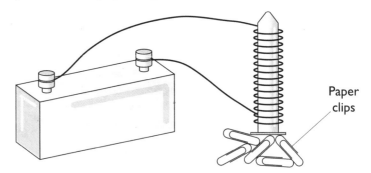

Paper clips

U SING ELECTROMAGNETISM

A useful electromagnet is made using a coil of wire on an iron core. This can cause movement when the magnet is switched on and off. Motors, bells and relays all use the movement caused by forces due to an electromagnet.

A **relay** is a switch operated by switching an electromagnet on and off. It may seem strange to use a switch to operate a switch, but using a relay does have advantages in certain situations.

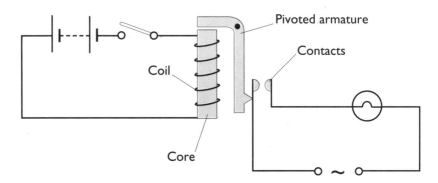

Pivoted armature

Contacts

Coil

Core

The diagram shows a relay being used to switch a mains lamp. The sequence of events when the relay is switched on is :

1 current passes in the relay coil, creating a magnetic field
2 the iron core becomes magnetized
3 the armature is attracted to the core
4 the switch contacts are pressed together and the light comes on.

Relays can operate from a low voltage source and be used to switch devices that work from a higher voltage.

Another advantage of the relay is that it only needs a small current to operate it. The electrical circuits in cars carry large currents that need thick wires. Relays are used so that thin wires can be used for the switches that the driver operates. This relay is used to switch car headlights.

Some outside lights have an infra-red sensor. The control circuit in this type of lamp uses a low voltage; it operates a relay to switch on the lamp when a person approaches.

Electric bells used in fire alarms rely on electromagnets. The electromagnet attracts the armature to make the bell 'ding'. The bell also incorporates a switching device called a 'make-and-break' so that when the bell has made a 'ding' the electromagnet is switched off, and then switched back on and off again repeatedly.

Now test yourself

7 The statements describe the events that happen when a bell push is operated.
Put them in the correct order.

A the circuit is broken at the contact screw

B the armature is attracted to the electromagnet

C the armature springs back

D the circuit is re-made and the process is repeated

E current passes in the electromagnet

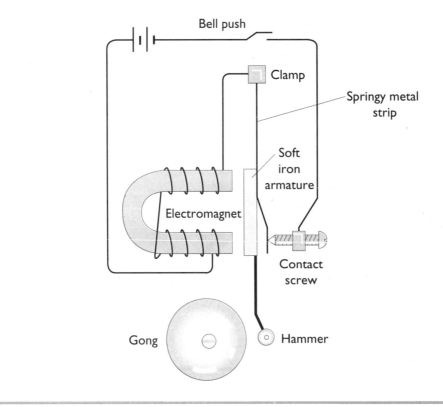

CHAPTER 11

*F*orce and motion

WHAT DO FORCES DO?

Forces are acting everywhere. No matter where you look, you will see evidence of things pushing and pulling other things.

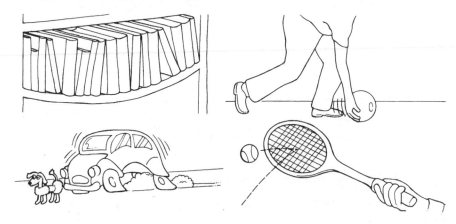

These drawings show some of the things that forces do. They start and stop things from moving, cause changes in direction, and change the shape of things when they squash or bend or stretch or twist them.

When you draw forces on a diagram, you should use an arrow to show the direction of the push or pull. Forces are described by a phrase such as 'object A pulls/pushes object B'. Here are some examples:

The Earth pulls the Moon   **The man pushes the pram**   **The girl pulls the door**

Now test yourself

1 Write a sentence to describe each of the forces shown in the diagrams below.

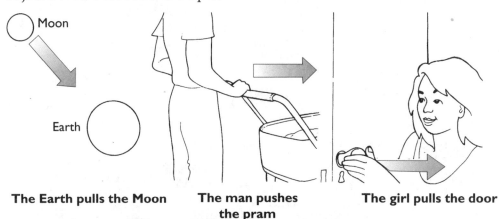

The first diagram in Question 1 shows the Earth pulling an apple. The Earth, like all other very massive objects, pulls everything else towards it. This force that pulls an object down towards the Earth is called its **weight**.

Like all other forces, weight is measured in **newtons** (N for short). On the Earth the weight of each kilogramme of material is about 10 N, so a 25 kg sack of potatoes is pulled towards the Earth with a force of 250 N.

The Moon also pulls things towards it but with a smaller force; on the Moon the potatoes would weigh about 38 N.

S TAYING STILL

Look around you; you are surrounded by things that are not moving. Perhaps you are seated at a desk that is not moving, or you may be sitting in a comfortable chair that is not moving. Even if you are reading this while seated on a moving bus, you can look out of the window and see things that are not moving.

Everything that you can see has at least one force acting on it – the Earth's pull. If something is not moving there must also be another force pushing or pulling it so that the forces are **balanced**.

When you sit on a chair it changes shape; the springs and cushion get squashed so that they push up on you. You now have two equal-sized forces acting on you in opposite directions – the forces on you are balanced.

Here are some more examples of balanced forces.

The downward pull of the Earth

The upward push of the shelf

The left-hand team's pull on the rope

The right-hand team's pull on the rope

If you push a plastic ball into a bucket of water you can feel the water pushing it back up – the more you push it in, the bigger the force from the water becomes.

Air also pushes up on things. It is the upward push of the air that causes a hydrogen-filled balloon to rise.

When a ball floats, the forces on it are balanced; the upward push of the water is equal in size to the downward pull of the Earth.

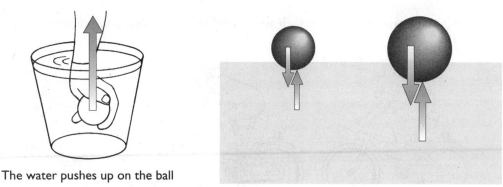

The water pushes up on the ball

A heavier ball has to displace more water because it needs a bigger upward force.

GETTING MOVING

A train, a car, a bus, a bike, a person walking; they all need a force to make them move. To start or stop movement, to speed up or slow down there needs to be an **unbalanced** force.

This unbalanced force makes the cyclist speed up

And this one slows him down

Can you identify the forces shown in the diagrams of the cyclist? The one on the left is the **driving** force that pushes the cycle along; that on the right is the **air resistance** that acts against the cyclist's motion. Anyone who has ever ridden a bicycle has felt the effect of air resistance; the faster you go, the bigger this resistive force gets as you have to push more air out of the way each second.

When you are riding a bicycle there are always resistive forces. Most of the resistance to motion comes from the air, but there are also resistive forces in places such as the wheel bearings that oppose a cyclist's motion. Because of these resistive forces, you have to keep pedalling just to maintain a steady speed. Putting the brakes on creates an extra resistive force so that you slow down more rapidly.

The driving force and the resistive force are both at work when a cyclist is pedalling. For the cyclist to speed up, the driving force needs to be bigger than the resistive force. If the resistive force is bigger than the driving force, the cyclist slows down. Cyclists usually stop pedalling when they brake, so there is no driving force, but only the resistive force acting!

When you get on a bicycle and start to pedal, the resistive force is small so you speed up quite rapidly. As your speed increases so does the resistive force. Eventually you get to a speed where the resistive force is equal in size to the driving force.

When this happens you stop speeding up and travel at a constant speed.

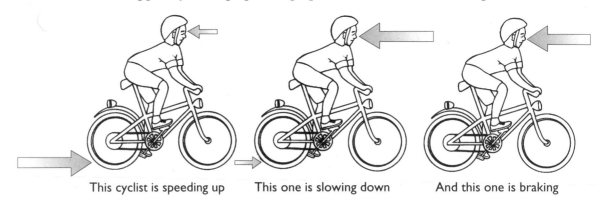

This cyclist is speeding up   This one is slowing down   And this one is braking

The forces acting on an object moving in a straight line at a constant speed, like those on a stationary object, are balanced. They are the same size but act in opposite directions so that their combined effect is just as if there were no force acting at all.

This cyclist is travelling at a constant speed

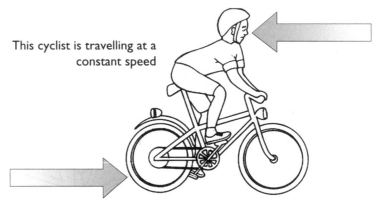

Unbalanced forces are needed to cause any change in motion. A change in direction needs an unbalanced force.

The diagram shows a girl on a trampoline. To make her move up when she bounces, the upward push of the trampoline has to be bigger than the downward pull of the Earth.

When the girl has lost contact with the trampoline and is moving upwards, the Earth's pull is unbalanced. This causes her to slow down and change direction, making her fall again.

Now test yourself

2 The diagram shows the horizontal forces on a car that is travelling at a steady speed.

(a) What does the arrow labelled A represent?

(b) What does the arrow labelled B represent?

(c) How can you tell that the car is travelling at a steady speed?

(d) Draw a diagram to show the forces on a car that is speeding up.

(e) Draw a diagram to show the forces on a car that is braking.

R ESISTIVE FORCES

When you swim, you have to work to push against the water. When you run, you have to work to push against the air. You can probably run faster than you can swim because the resistive force from the air is less than that from the water.

There are other resistive forces that act on moving objects. Sometimes their effect is to slow down or stop the motion, but they also make motion possible. Without resistive forces we cannot walk, and bikes, cars, buses and trains cannot move.

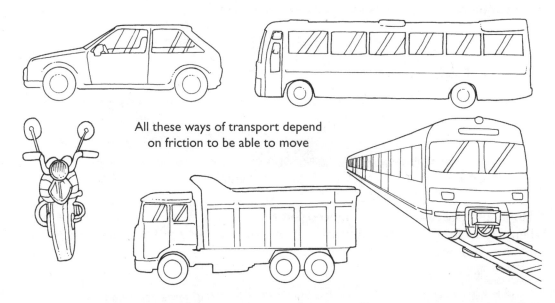

All these ways of transport depend on friction to be able to move

Friction is a resistive force that acts against sliding or slipping. If you push a book across a desk, friction is the force that slows it down and stops it. Friction always acts in the opposite direction to any sliding motion.

Friction force

Motion of book

Rough surfaces cause bigger friction forces than smooth ones do. So if you want something to slide, you should keep the surface smooth and polished or lubricated. Ice is a good surface to slide on but a very poor surface to walk or ride a bike on.

When we walk, we rely on friction to stop our feet from slipping. To walk forwards our feet push backwards on the ground. The friction force stops them from moving backwards and pushes us forward. Without friction, our feet would just slip and we would not get anywhere!

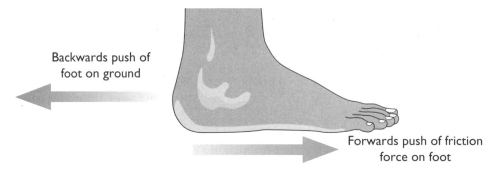

Backwards push of foot on ground

Forwards push of friction force on foot

Trains also need friction between the wheels and the rails. When wet leaves get on the track the friction force is reduced and the train wheels slip.

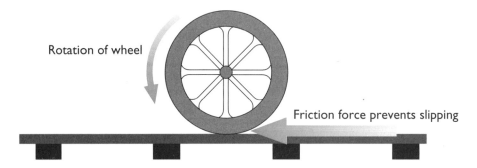

Rotation of wheel

Friction force prevents slipping

Friction also provides the unbalanced force for a car or bicycle to change direction when it goes round a corner. In this case the force acts at right angles to the direction of motion.

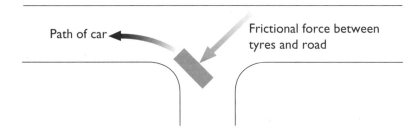

Path of car

Frictional force between tyres and road

Parachutists depend on resistive forces to slow them down. A sky-diver who jumps from an aircraft speeds up to a speed of about 60 m/s. At this speed the air resistance balances the Earth's pull on the sky-diver.

Opening the parachute causes the air resistance to get bigger, so the forces acting on the sky-diver are no longer balanced. The sky-diver now slows down.

Eventually the sky-diver falls at a much lower speed with the Earth's pull and the air resistance once again in balance.

Now test yourself

3 The graph shows how the speed of a parachutist changes after she has jumped from an aircraft.

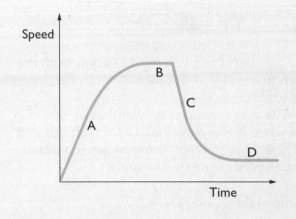

(a) Describe what is happening to the speed of the parachutist during part A of the graph.

(b) During which parts of the graph are the forces on the parachutist balanced? Explain how you can tell.

(c) During which part of the graph is the weight of the parachutist bigger than the air resistance? Explain how you can tell.

(d) Describe and explain what is happening to the speed of the parachutist during part C of the graph.

SIR ISAAC NEWTON (1642–1727)

Sir Isaac Newton was born in 1642, the son of a Lincolnshire farmer who died when Isaac was three years old. His mother re-married and left Isaac in the care of his grandmother who was instructed to bring him up to be a farmer, like his father. This did not suit Isaac and his uncle persuaded his mother to send him to grammar school in Grantham. From there, in 1661, he went to study at Trinity College, Cambridge.

After he obtained his degree in 1665, the University was closed for two years because of the plague. Isaac continued his studies himself, working on a branch of mathematics that is now called calculus. He returned to Cambridge in 1667 and was made Professor of Mathematics in 1669.

Isaac continued to work very much on his own.

He was interested in optics and colour and he was the first person to realize how white light can be split into different colours. After a visit from Edmund Halley, a famous mathematician and astronomer, in 1684, Isaac pursued his study of motion and gravitation.

The result of these studies was his three laws of motion and his law of universal gravitation; some of the greatest discoveries in the history of science. These laws are used every day by scientists all over the world to make predictions about the movement of objects.

Isaac Newton's life was not without conflict; in particular there were quarrels with Robert Hooke, who claimed that Newton had stolen one of his ideas, and the German mathematician Gottfried Leibniz. Leibniz complained about Newton to the Royal Society. Newton was president of the Royal Society; he set up a committee to investigate the complaint and he secretly wrote the report that found against Leibniz and in his own favour.

After leaving Cambridge, Isaac Newton moved to London where he became Master of the Royal Mint.

H OW FAST?

How fast can you run, cycle, swim? Who is the fastest person in your class? Which is the fastest way of crossing the English Channel: by ferry, hovercraft or train?

To answer these questions you need to compare measurements. Sometimes only one measurement is needed; you can tell which is the fastest method across the English Channel by comparing journey times.

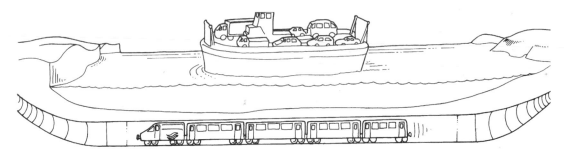

If you want to work out the **speed** of a moving object you need to know both the **distance** it travelled and the **time taken** to travel that distance.

If a bus travels 60 miles in 2 hours then its average speed is 30 mph. This speed is only an average; there would be times when it was travelling faster than this and times when it was travelling slower. There were probably times when it was not moving at all. You can work out average speeds using the formula:

average speed = distance travelled ÷ time taken   or $s = \dfrac{d}{t}$

Here is an example:

Calculate the average speed of a train that travels the 180 miles from York to London in 2 hours.

Answer: average speed = distance travelled ÷ time taken
                          = 180 miles ÷ 2 hours
                          = 90 mph

Although we use miles and hours when talking about everyday journeys, in science distance is measured in metres and time is measured in seconds. The next example uses scientific units:

Calculate the average speed of a motorcycle that takes 6 seconds to travel 90 metres.

Answer: $s = \dfrac{d}{t}$

            $= \dfrac{90 \text{ m}}{6 \text{ s}}$

            $= 15 \text{ m/s}.$

You need to know what your average speed is likely to be if you want to estimate the time for a journey. In good weather conditions, provided there are no delays such as traffic jams, the average speed of a car travelling on a motorway is about 60 mph. How long would a journey of 150 miles take?

"Dad, are we nearly there?"

To answer this question, you need to be able to use the speed equation in the form 'time = ...'.

The speed equation can be written down in three different ways. In symbols, these are:

$$s = \frac{d}{t} \quad t = \frac{d}{s} \quad \text{and} \quad d = s \times t$$

These are all different ways of writing the same equation.

You can now work out the journey time using $t = \frac{d}{s}$. You should get 2½ hours.

Now test yourself

4 Work out the times, distances and speeds that go in the blank spaces in the table. Take care to write the correct unit with your answer.

	Distance travelled	Time taken	Average speed
(a)	300 m	6 s	
(b)	6 cm	1.5 s	
(c)		5 hours	125 km per h
(d)	1750 miles		500 mph
(e)	7.5 m	0.5 s	
(f)	1500 m		25 m/s
(g)		4.5 s	8 m/s

U SING GRAPHS

Line graphs are often used to show an object's motion. A **distance–time** graph shows how the total distance travelled changes with time. The distance travelled by an object can only increase with time. A **speed–time** graph shows how the speed changes with time. Speed can increase and decrease so the line can go down as well as up.

The graphs shown below are for the same journey.

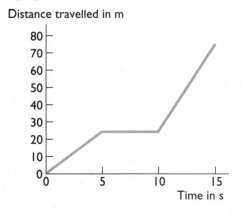

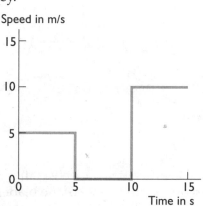

Both graphs show a constant speed of 5 m/s for the first 5 s, followed by a 5 s rest and then a constant speed of 10 m/s for the last 5 s shown.

Now test yourself

5 A car is travelling along a road at 30 m/s. The car driver sees some traffic lights at red and brakes to a halt. The graph shows how the speed of the car changes from the moment that the driver sees the traffic lights.

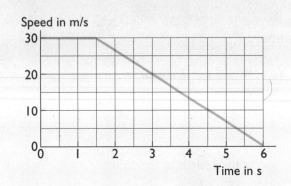

(a) How far did the car travel before the brakes were applied?

(b) After the brakes were applied, how long did it take for the car to stop?

(c) The average speed when the car was braking was 15 m/s. Calculate the distance the car travelled after the brakes were applied.

(d) What was the total stopping distance for the car?

U NDER PRESSURE

Pressure describes the effect a force has in piercing or deforming the surface it acts on. Knives, scissors, needles and drawing pins are all designed to cut or pierce; they create a large pressure by applying the force onto a small area.

Skis and caterpillar tracks on heavy vehicles are examples of spreading a force over a large area to reduce the pressure it causes.

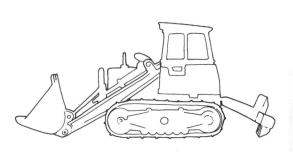

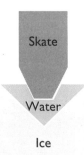

Ice skaters do not skate on ice; they skate on water. An ice skater's weight pushes down on the small area of the blades. The high pressure on the ice below the blades causes it to melt, so the blades are surrounded by a film of water. The water re-freezes when the ice skate has passed. This is how an ice skater leaves tracks in the ice – the tracks are where the ice has melted and then re-frozen.

Now test yourself

6 Explain the following:

(a) It is easier to cut cheese with a wire than with a knife.

(b) People are not allowed on bowling greens unless they are wearing flat-heeled shoes.

(c) It is more comfortable to sit on a soft cushion than on a hard stool.

QUANTIFYING PRESSURE

Pressure is calculated as the force acting on each cm^2 or m^2 of surface area using the formula

$$\text{pressure} = \frac{\text{force}}{\text{area}}$$

It is measured in N/cm^2 or N/m^2. An alternative unit for N/m^2 is the pascal (Pa for short).

Here is an example:

A bulldozer weighs 150 000 N. To stop it from sinking into the soft mud it moves on caterpillar tracks. The area of the tracks in contact with the ground is 10 m^2. Calculate the pressure on the ground.

$$\begin{aligned}
\text{Answer:} \quad \text{pressure} &= \frac{\text{force}}{\text{area}} \\
&= \frac{150\,000 \text{ N}}{10 \text{ m}^2} \\
&= 15\,000 \text{ N/m}^2 \text{ or } 15\,000 \text{ Pa.}
\end{aligned}$$

Like the speed equation, the pressure equation can be written in three different ways:

$$P = \frac{F}{A} \quad F = P \times A \quad \text{or} \quad A = \frac{F}{P}$$

Now test yourself

7 Use the pressure equation to fill in the blanks in the table. Take care to write down the correct unit in each case.

	Force	Area	Pressure
(a)	25 N	2 m^2	
(b)		0.5 m^2	500 Pa
(c)	50 N		500 Pa
(d)	100 N	0.1 cm^2	
(e)		3 m^2	100 000 Pa

T URNING FORCES

Whenever we turn on a tap, push the pedals on a bike or open a door we are using a force to turn something round. The point that things turn around is called the **pivot**.

In the case of a tap, the pivot is at the centre of the tap; a door pivots around the hinge. The diagram shows the force and the pivot when a pedal is turned.

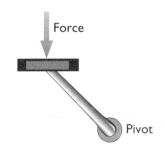

If you try closing a door by pushing it at different places, you realize why door handles are placed as far from the pivot as possible.

The effect that a force has in turning something round is called the **moment** of the force. It depends on the size of the force and how far away from the pivot it is applied; the further away from the pivot, the bigger the turning effect of a force.

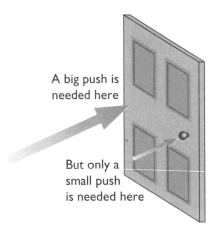

A big push is needed here

But only a small push is needed here

The moment of a force can be calculated using the equation:

moment = force × perpendicular distance to pivot

The diagrams show how this equation is used to calculate the moment of a force used to open a door.

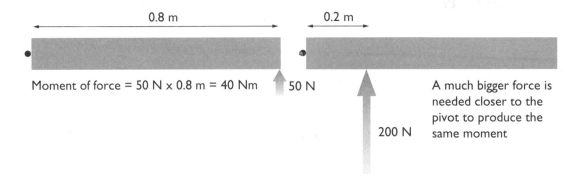

0.8 m

0.2 m

Moment of force = 50 N x 0.8 m = 40 Nm

50 N

200 N

A much bigger force is needed closer to the pivot to produce the same moment

Now test yourself

8 Each of the diagrams below shows something that can be made to turn round. On each diagram, mark the pivot and a force that would cause turning.

9 Calculate the moment of the force being used to tighten the wheelnut.

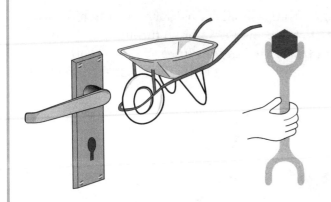

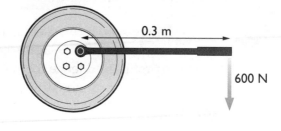

0.3 m

600 N

A QUESTION OF BALANCE

A see-saw has two forces acting on it. Each of them has a turning effect, but they are acting in opposition. If the moments of the forces are unequal, then the one with the bigger moment wins, but if the forces have equal moments the see-saw is balanced.

This is an example of the **principle of moments**, which states that 'if an object is balanced, the sum of the clockwise moments about a pivot is equal to the sum of the anticlockwise moments about the same pivot'. This sounds complicated, but in fact it is easy to apply.

Now test yourself

10 The diagrams show some see-saws. For diagrams A, B, C and D, decide whether each one is balanced, or whether it will rotate clockwise or anticlockwise. For diagrams E and F work out the size of the force needed to balance the see-saw.

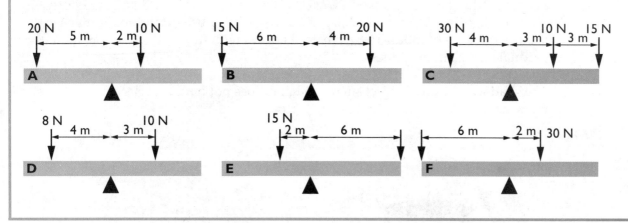

CHAPTER 12

Light and sound

SHADOWS

Sharp shadows are formed when light from a small source such as a torch passes around an object that is **opaque**, i.e. that does not let light go through it. These shadows provide evidence that **light travels in straight lines**. If light could travel in curved lines or round corners some light would get into the black shadow region shown in the diagram.

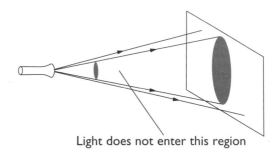

Light does not enter this region

Light from the Sun also forms sharp shadows. Although the Sun is not a small light source, its distance from the Earth is so great that it behaves like one.

Sundials, a way of telling the time of day before clocks were invented, used the movement of the Sun's shadow on a dial. You can see the same effect on a sunny day by watching the shadow of a stick placed in the ground.

Although a shadow caused by the Sun is sharp, it is not totally black because light that has been scattered from the sky can enter the shadow region. Shadows on the Moon and planets with no atmosphere are black.

People working in a kitchen or a workshop where attention to detail is required do not want lighting that creates sharp shadows, so they use **extended light sources** such as fluorescent tubes. These create shadows with fuzzy edges, where there is a gradual change in brightness.

The shadow of a small object such as a hand does not have a dark centre. It is just a fuzzy area that is not as bright as the surroundings.

Outside the dark region, the brightness gradually increases

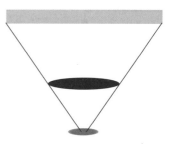

Some light can reach all parts of the shadow of a small object

Now test yourself

1 When there is an eclipse of the Sun, part of the Earth is in the Moon's shadow. The diagram shows the Earth, Sun and Moon during an eclipse.

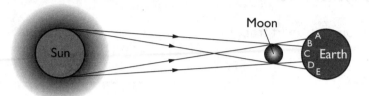

On which of the lettered part or parts of the Earth:

(a) is there no sunlight?

(b) is the eclipse not visible?

(c) does the light level vary from bright to dark?

H OW WE SEE

Some people think that we see with our eyes, but we also need a brain to be able to see things! Our eyes are sensors that send messages to the brain in the form of electrical signals along the optic nerve. When it interprets the signals from the eyes, the brain assumes that the light travelled to the eyes in a straight line. This assumption enables the brain to work out where things are. It needs light from both eyes to be able to pinpoint the position of things accurately.

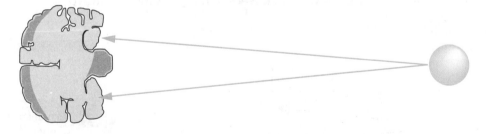

Television and computer screens are **luminous**: they give out light that our eyes detect. Lamps and the Sun also give out light but we do not often look directly at these. Looking at a lamp can cause temporary blurred vision and you should **never** look directly at the Sun as this could cause permanent eye damage.

We use light sources such as the Sun and artificial lighting so that we can see other things. Most surfaces **scatter** light, that is they **reflect** it in all directions, so that some light enters your eyes even when you move to a different position.

Now test yourself

2 Name **five** luminous objects, apart from the ones mentioned in the text.

3 You are sitting in a chair after dark, using a reading lamp to help you to read a book. Describe how you can see the book.

M IRRORS

Unlike most everyday things, mirrors do not reflect light in all directions. But they do reflect it in a regular and predictable way. Light is reflected from a mirror at the same angle as it hits it.

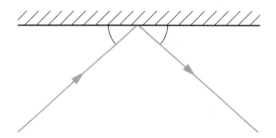

We use this regular reflection when we look at our image in a mirror. Looking into a mirror causes our brain to get confused: it 'sees' things that aren't really there!

Light from your nose (and other parts of you!) hits the mirror and is reflected at equal angles. The reflected light is detected by your eyes. Your brain then 'sees' the nose, and works out where it is assuming that the light has travelled in straight lines. In this case it gets things wrong: it 'sees' a nose behind the mirror.

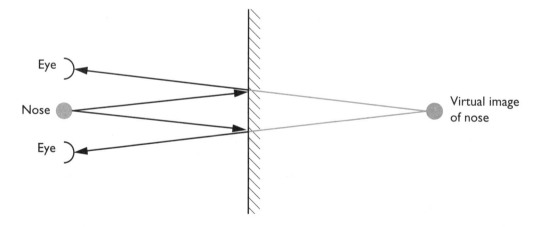

What you see is called a virtual image. The word 'image' just means 'likeness', and in all respects the image is a likeness of the real thing; it is the same way up and the same size and colour. The image is called virtual because, unlike the image that you see on a television or cinema screen, it is not really there.

Using the rule about the way in which light is reflected at a mirror, you can construct images. If you do this, you will see that the image in a mirror is always formed straight behind the mirror, the same distance behind it as the object is in front of the mirror.

Mirrors are used in periscopes to turn light round corners. Light hitting a mirror at an angle of 45° is reflected at the same angle and so is effectively turned through 90°.

The diagram shows how two mirrors are arranged inside a periscope. You can make a periscope quite easily using two small mirror tiles and the cardboard tube from the inside of a roll of kitchen foil.

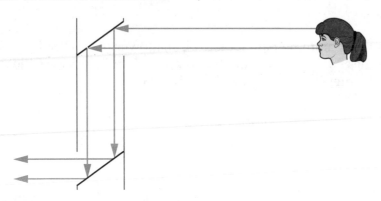

Now test yourself

4 Which diagram shows the reflection of light by a mirror correctly?

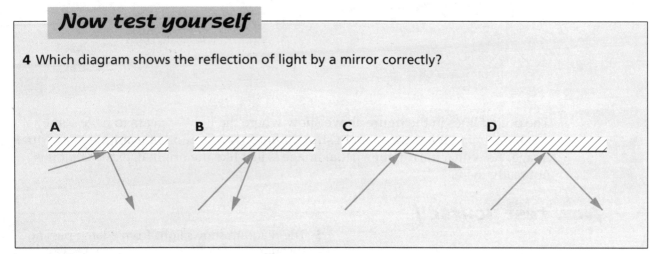

A   B   C   D

MORE IMAGES

Have you ever noticed how the water in a swimming bath never looks to be as deep as it really is? This is another example of your eye–brain system being fooled when light does not travel in straight lines. A change in the speed at which light travels can also cause a change in direction.

Light travels in the vacuum between the Earth and the Sun at a speed of 300 000 000 m/s. It takes just over eight minutes to reach the Earth. As it enters the Earth's atmosphere the light is slowed down a little and continues to slow down as the atmosphere gets denser. However, the speed is changed very little by the atmosphere and so we still use the figure of 300 000 000 m/s when describing the speed of light in air.

Light that enters water or any other dense transparent material is slowed down much more. In water, light travels at approximately 230 000 000 m/s and in glass at 200 000 000 m/s. This change is speed is called **refraction**. The diagrams show the effect of the change in speed when light passes through a block of glass.

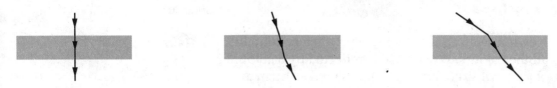

The light that meets the boundary between the air and the glass at an angle of 90° carries on without a change in direction. At any other angle the light changes direction as it goes into and leaves the glass block.

Light travelling through water undergoes a similar change in direction. The amount of change is slightly less because there is less reduction in speed as light enters water.

So why does the swimming bath look to be shallower than it really is? The answer is because of the change in direction that occurs when light leaves the water and speeds up as it enters the air.

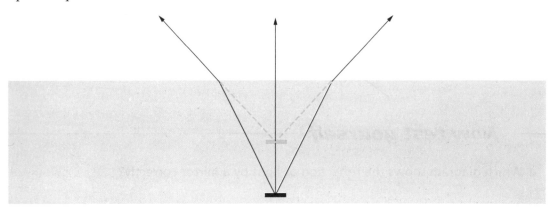

The dotted lines in the figure above show where the light appears to have come from, assuming it has travelled in straight lines. This is another example of a virtual image. As with a mirror, the virtual image is just like the original, except that it is not really there.

Now test yourself

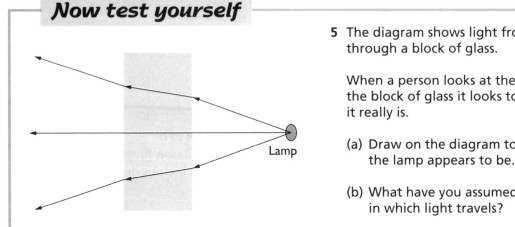

Lamp

5 The diagram shows light from a lamp passing through a block of glass.

When a person looks at the lamp through the block of glass it looks to be nearer than it really is.

(a) Draw on the diagram to show where the lamp appears to be.

(b) What have you assumed about the way in which light travels?

COLOUR

Manufacturers of paint make different colours and shades by starting with a white paint and then adding dyes to it. Until the time of Isaac Newton, it was thought that coloured objects did the same thing to light.

Newton realized that white light already contains all the colours. When light is reflected by a coloured object, instead of adding colours it takes them away.

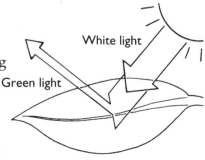

White light

Green light

Newton made his discovery when he noticed that white light going into a glass prism emerged split into different colours. He found that he could make these colours combine together again to make white.

You may have seen white light split into colours when it passes through a triangular prism. This is called **dispersion**. It is caused by the different colours of light travelling at different speeds in the prism material. Light is also dispersed into colours by water droplets, causing a rainbow.

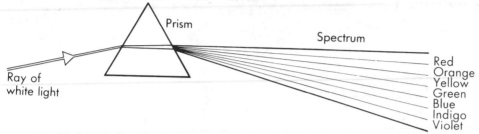

When we watch television we are seeing the effects of combining colours together. A television screen can glow with three different colours: red, green and blue.

These are called the **primary** colours because all the other colours in the spectrum can be made by combinations of these three colours.

The diagram shows how the three **secondary** colours, yellow, magenta and cyan (turquoise), are each produced by mixing the light from two primary colours.

The area in the middle, where all three primary coloured lights mix, is white.

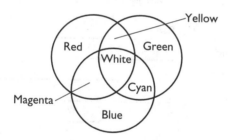

Now test yourself

6 Write down which two primary colours, from red, green and blue, are added together to make each secondary colour.

COLOUR SUBTRACTION

Televisions and computer screens produce all the colours in the rainbow by adding the three primary colours together in varying proportions. Coloured objects and colour filters take colours out of the light that falls on them.

Primary colours take out everything except their own colours.

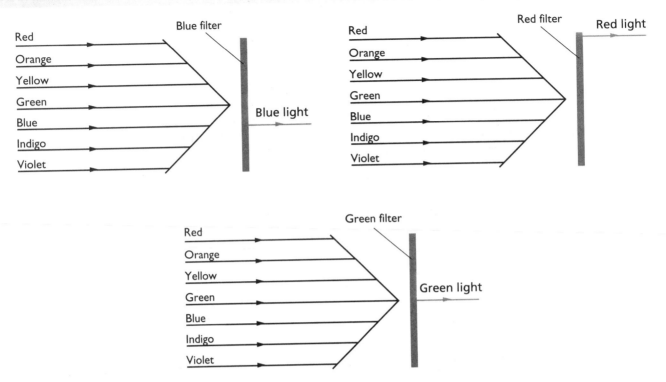

Each secondary colour removes just one primary colour – the one that isn't used when mixing the secondary colour from red, blue and green light. So cyan removes red light, yellow removes blue light and magenta takes away green light.

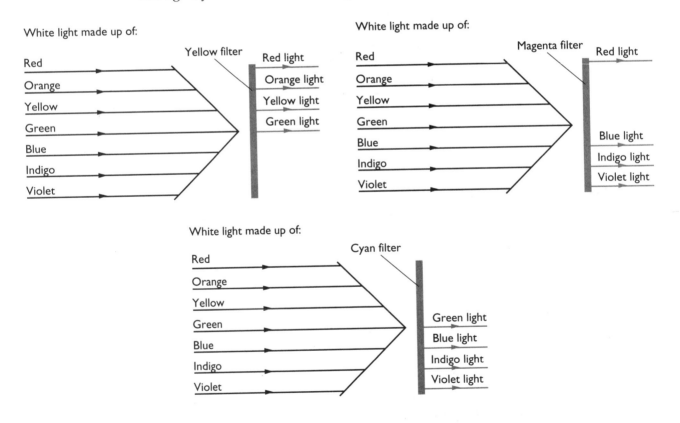

If you go to a disco where there are different colours of light, you will notice that clothes seem to take on different colours. A cyan-coloured T-shirt looks black in red lighting and green in yellow lighting.

To work out what colour different objects look under coloured lights, start with the primary colours present in the light. Then take away the colours that the object absorbs, and you are left with the colours reflected. It's just a simple subtraction sum.

For example, a magenta-coloured light shines onto a yellow shirt. What colour does the shirt look to be?

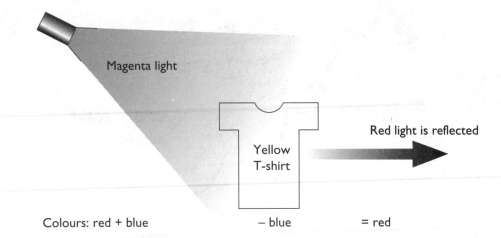

Magenta light

Yellow T-shirt

Red light is reflected

Colours: red + blue   − blue   = red

Sometimes an object takes away all the colours in the light. When this happens it appears black.

Now test yourself

7 A boy wears a sweater which is turquoise (cyan) when viewed in daylight.

(a) Which primary colour is absorbed by the turquoise sweater?

(b) What colour would the sweater look to be when viewed in the following colours of light:
(i) yellow?
(ii) red?
(iii) blue?
(iv) magenta?

8 White light passes through a yellow filter and then a cyan filter. What colour emerges from the cyan filter?

SOUND

To make a sound, something has to vibrate in a to-and-fro motion. Televisions, radios and hi-fi all use loudspeakers to reproduce sound. A loudspeaker is a paper cone driven backwards and forwards by an electromagnet.

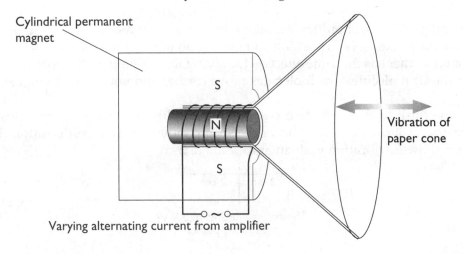

Cylindrical permanent magnet

S

N

S

Vibration of paper cone

Varying alternating current from amplifier

Some musical instruments have strings that vibrate when they are hit or plucked; others have columns of air that vibrate when they are blown.

Sound can travel through anything that has particles capable of transmitting the vibrations. It travels very fast in solids, where the particles are close together, slower in liquids and slowest of all in gases.

You may have seen a demonstration of a bell ringing inside a glass jar. Sound from the bell travels through the air and then glass and then air again to the ear. When air is removed from the jar, no sound can be heard.

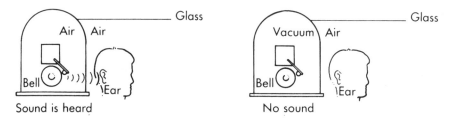

A slinky spring can be used to show how sound waves travel through materials. As the wave moves along the spring, each part of it vibrates. The molecules in air vibrate in a similar way when they transmit sound. Each wave consists of a compression or 'squash' where the molecules are close together, followed by a rarefaction or 'stretch', where the molecules are further apart.

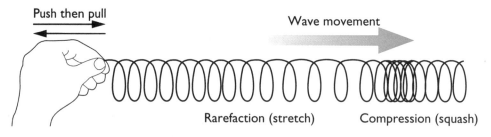

Your hand pushing and pulling a slinky spring is a good model of what happens when a loudspeaker cone is pushing and pulling on the air. Using this model, moving your hand further has the same effect on the spring that turning up the volume control has on the air molecules: the molecules move further and you hear a louder sound.

When you push further you are increasing the **amplitude** of the vibration; this is the greatest distance that each part of the spring moves from its rest position. The diagram shows a vibration with an amplitude of 2 cm.

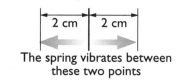

A microphone and an oscilloscope can be used to plot a graph that shows the movement of air molecules when a sound wave passes. The upwards and downwards movement of the oscilloscope trace represents the forwards and backwards movement of an air molecule.

The distance marked with an arrow represents the amplitude of the wave. A louder sound of the same pitch has a larger amplitude.

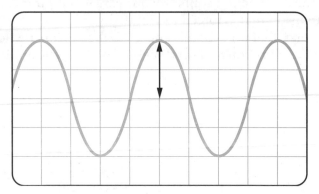

The trace on an oscilloscope screen shows the number of sound waves that are detected in a certain time; the actual time can be changed by adjusting the oscilloscope settings. One complete wave is shown by a 'crest' and a 'trough'. The trace shown in the diagram above shows two and a half waves. The next diagram shows the trace when a note of higher pitch is played.

There are now more waves in the same time; the **frequency** (the number of waves per second) has increased.

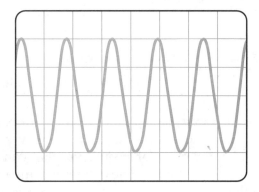

These measurements using an oscilloscope display show that:

1 An increase in **amplitude** of a sound wave makes it sound louder.
2 An increase in **frequency** of a sound wave gives it a higher pitch.

Now test yourself

9 You can make a sound by plucking a stretched rubber band.

(a) How does the rubber band make a sound when it is plucked?

(b) Describe how the sound is carried from the rubber band to your ears.

(c) How does the sound change if the rubber band is plucked harder? What causes this change?

(d) If the rubber band is stretched more, when it is plucked the sound has a higher pitch. What change has caused the higher pitch?

H EARING

Sound reaches our ears as the energy of vibrating air molecules. The vibrations have a tiny amplitude, typically a few thousandths of a millimetre, but our ears are very sensitive to small amounts of energy transmitted in this way.

The first stage in the detection of sound by our ears is the **ear drum**. This is a thin sheet of muscle and skin that is forced to vibrate by the air in the outer ear. The ossicles are three bones that transfer this energy to the inner ear where nerve endings are stimulated by the vibration and send messages along the auditory nerve to the brain.

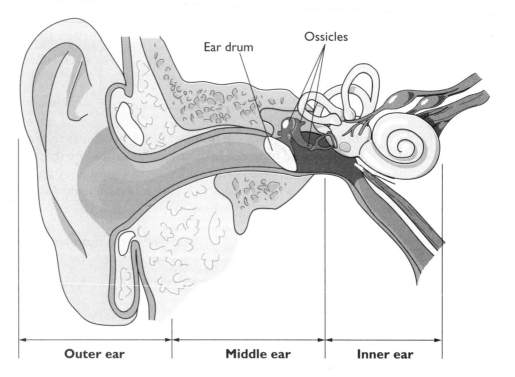

Not everybody can hear the same sounds! The ears of a normal young person detect sounds with a frequency ranging from about 20Hz (20 waves per second) to about 20 000 Hz. As you get older, the range of frequencies that you can hear is reduced. A middle-aged person may not be able to hear sounds with a frequency greater than 15 000 Hz, although the hearing at low frequencies is less likely to be affected.

Even within this range, your ears are more sensitive to some frequencies than to others. Sounds at a frequency of around 2000 Hz sound louder than higher or lower sounds.

Other animals have different frequency ranges. It is well-known that bats can emit and detect sound waves that are well above the range of human hearing, and dog whistles use a high frequency that we cannot hear. Elephants and dolphins can communicate over long distances using very low frequencies that have a longer distance range than high-frequency waves.

People who work in noisy environments should wear ear muffs to protect their ears from damage. Sudden exposure to a loud sound such as an explosion can cause immediate damage by breaking the ear drum or the ossicles.

The ossicles are also subject to wear; they are pieces of machinery in constant use. A person who is repeatedly subjected to loud sounds, such as those from pneumatic drills or discos, suffers loss of hearing as the ossicles wear away. Surgeons can sometimes replace worn out ossicles with plastic ones to improve a person's hearing.

Now test yourself

10 Which two of the following frequencies of sound waves cannot be detected by most human beings?

10 Hz   35 Hz   1000 Hz   15 000 Hz   25 000 Hz

11 Disc jockeys and pop singers often suffer from hearing loss at a young age. Suggest what causes this.

CHAPTER 13

*T*he Earth and beyond

T HE VIEW FROM EARTH

If you are sitting at home reading this, you do not have a sensation of movement. It is not obvious that our planet Earth is moving through space with a speed of thousands of metres per second. So it is not surprising that early astronomers made the mistake of thinking that the Sun moves around the Earth.

On a sunny day you can plot the Sun's path across the sky by watching the movement of a shadow.

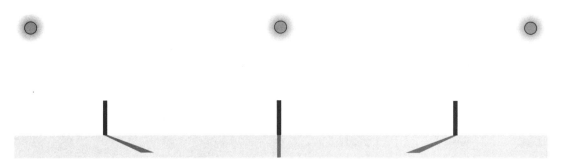

The shadow is longest in the morning and evening when the Sun is low in the sky. As the shadow moves round in the morning it gets shorter, and is at its shortest at noon when the Sun has reached its highest point in the sky. After noon, the shadow starts to lenghthen again. You can see very long shadows near sunset on a sunny evening.

These changes in a shadow seem to be caused by the Sun moving across the sky – rising in the East and setting in the West. The diagram shows how the Sun appears to move each day.

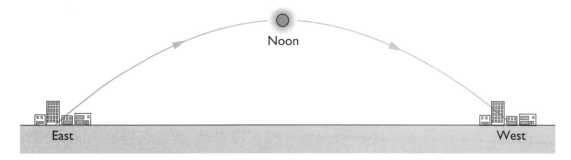

We now know that this 'movement' of the Sun is actually due to the Earth turning round once each day. The Earth spins around its axis, an imaginary line going through the centre of the Earth from pole to pole.

NICOLAUS COPERNICUS (1473–1543)

Copernicus was a Polish astronomer who first put forward the theory that the Earth goes round the Sun. This theory was rejected at the time when it was widely believed that the Earth was the centre of the Universe, with the celestial bodies revolving round it.

Copernicus was born in Torun to a wealthy family. After studying liberal arts at the University of Krakow, he went to Italy to study medicine and law. He became interested in astronomy while he was studying at the University of Bologna.

After returning to Poland, Copernicus developed his theory that the Earth rotates once each day on its own axis and orbits the Sun once each year. He was also able to explain the apparent daily motion of the Sun and the stars and why some planets sometimes appear to go backwards.

Copernicus' ideas were rejected by all but a few astronomers. It was not until the time of Isaac Newton that his theories became widely accepted throughout Europe.

If you were out in space, looking down at the Earth's north pole, you would see the Earth turning round in an anticlockwise direction.

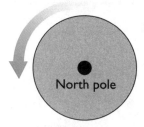

This daily rotation of the Earth causes day and night and the Sun's apparent movement across the sky. The diagram below shows Britain at sunrise, noon and sunset. This is the view you would have if you were looking down at the North pole. The outer circle of the Earth is the equator.

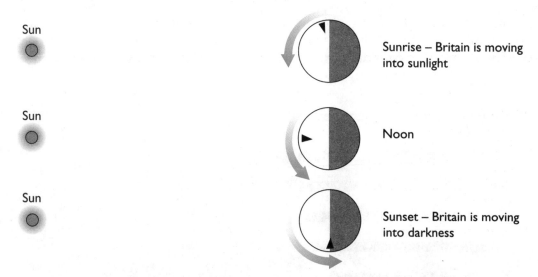

Sun — Sunrise – Britain is moving into sunlight

Sun — Noon

Sun — Sunset – Britain is moving into darkness

Use the diagram to imagine where you would have to look to see the Sun at different times of day.

Days vary in length in summer and winter, and the highest position of the Sun in the sky also changes with the season. Just twice a year, at the spring and autumn equinox, day and night are of equal length. Summer brings longer days than nights, with the Sun being higher in the sky. The diagram shows the Sun's apparent path at three different times of year.

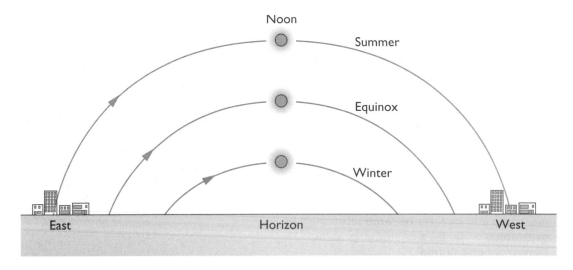

Different lengths of day and night are caused by the Earth being tilted. The Earth's axis always points in the same direction, towards the star Polaris.

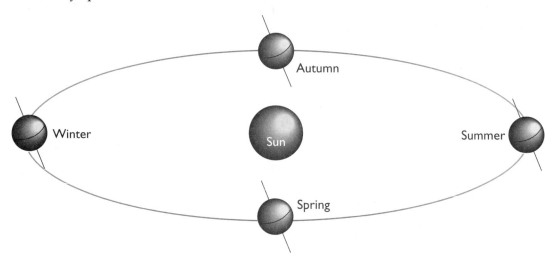

When it is summer in Britain the northern hemisphere is tilted towards the Sun, so we spend more than twelve hours in daylight. The nearer you are to the North pole in summer, the longer your days are. (The pole itself is light for twenty-four hours a day in the height of summer.)

Summer in the northern hemisphere

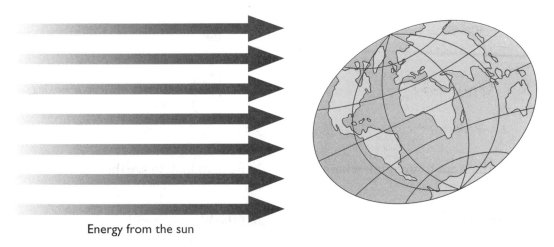

Energy from the sun

This not only makes the days longer, it also means that energy from the Sun is spread over a smaller area, giving us a warmer climate in summer.

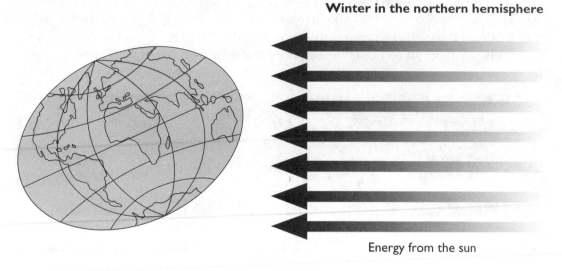

Winter in the northern hemisphere

Energy from the sun

When it is winter in Britain, energy from the Sun is spread over a large area. You can see from the figure that it is Australia's summer; energy from the Sun is not spread out over a large area.

Our Sun is the only star that we see during the day; the other stars are there but the bright light from the Sun prevents us from seeing them.

Polaris, or the Pole Star, is a very bright star. As mentioned earlier, the Earth's axis points towards Polaris, so you can see it by looking in the sky directly north. The diagram shows the Pole Star as it can appear in the winter sky, together with the constellations of the Plough and Cassiopeia.

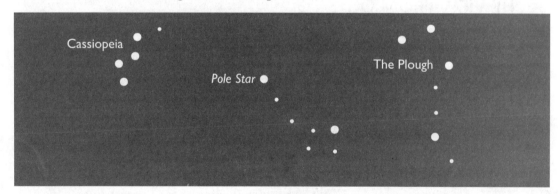

Cassiopeia

The Plough

Pole Star

The Pole Star is the one star in the sky that you can always see in the same place. You can see the effect of the Earth's rotation if you stand under the ceiling light in the centre of a room. Imagine that this is the Pole Star; the Plough is to your right. Now walk around the light anticlockwise through a quarter of a circle; the Plough now appears to be 'above' the Pole Star.

The Earth's spin on its axis makes the stars in the northern sky appear to revolve anticlockwise around the Pole Star. Stars in the southern sky appear to move round clockwise.

If the Earth's only movement was its spinning, we would see the stars in the same place at the same time each night. But we have to take into account the movement around the Sun. As we revolve once around the Sun it looks as if the stars are turning once around the Pole Star. Since there are 365 days in a year and 360° in a circle, the pattern of the stars seems to move by about 1° each day. So if you look at the stars at the same time of night, in one month they should have moved round in the sky by about 30°.

In winter, the constellation Orion can be seen in the southern sky. The left-hand diagram below shows what it looks like one winter's night at 9 pm. In the right-hand diagram it has moved round through 30°. 30° is one-twelfth of a rotation, so the right-hand diagram could have been drawn at 11 pm on the same night or at 9 pm one month later.

Now test yourself

1 The diagram, shows the Sun at midday in winter.

(a) Draw a line on the diagram to show the path that the Sun follows between sunrise and sunset.

(b) Write an **S** at the place where you would expect to see the Sun at midday in summer.

(c) What causes the apparent movement of the Sun across the sky each day?

THE VIEW FROM OUTSIDE

If you could look at the solar system from a viewpoint where you could see all the planets, what would you see? The most striking object would be the Sun, shining with a brilliant white light. You would also be struck by the rich colours of the planets and the contrast between the rich orange colour of Venus and the blue planet Earth. Of course, you would only be able to see half of each planet, because planets are only visible by the Sunlight that they reflect and only the half of each planet facing the Sun would be lit up.

GALILEO GALILEI (1564–1642)

Galileo is most famous for his astronomical discoveries using a telescope and for demonstrating that the speed of a falling object does not depend on its weight. His pursuit of science brought him into conflict with the Roman Catholic Church, which was only resolved in 1992, 350 years after his death.

Galileo was born near Pisa. After being taught by monks he studied medicine at the University of Pisa but he became more interested in science and mathematics.

In 1589, as Professor of Mathematics at Pisa University, Galileo is reported to have dropped two objects of different weights from the top of the leaning tower, showing that the long-held belief that heavy things fall faster than light things was wrong. The result of this was that he lost his job

at Pisa and he moved to the University of Padua.

In 1609 Galileo built a telescope that enabled him to observe the Moon, the Milky Way, and to discover four of Jupiter's moons. He supported the Copernican theory that the Earth moves around the Sun. It was this support for Copernicus that caused the Church to denounce him as a heretic. He was threatened with torture and forced to sign a document saying he did not believe that the Sun was the centre of the Universe. After a long dispute with the Church, Galileo was put under permanent house arrest in 1633.

As Galileo's life ended, Newton's was about to begin. Galileo's studies of motion provided the foundation for Newton to develop his theory of gravitation.

You would notice four planets close to the Sun. Nearest to the Sun is the tiny planet Mercury, its surface covered in impact craters.

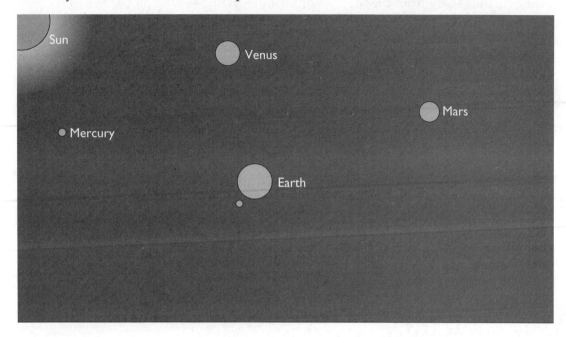

Next comes Venus, its dense atmosphere reflecting only orange light. The third planet out is the planet Earth, but from outer space you would not be able to see any evidence of human activity. You would notice that the Earth has a satellite of its own, a moon very similar in size and appearance to Mercury. The Earth's moon takes about twenty-eight days to complete an orbit of the Earth. It takes the same time to spin once on its axis, so the same side of the Moon always faces towards the Earth.

The fourth planet out from the Sun, Mars, appears a blood-red colour and you would certainly notice the similarities between the appearance of Mars and that of Earth. They have similar length days, they both have white clouds and they are both tilted so they have similar seasons.

If you concentrated on the four inner planets, you would see that they all go round the Sun in the same direction. You would also notice that the speed of the planets depends on the distance from the Sun. Mercury moves fastest of all; it also has the shortest distance to travel to complete an orbit, so it races round the Sun, completing five orbits to the Earth's one.

Moving your eyes further out, you might notice a lot of rocky fragments orbiting the Sun between Mars and the first of the outer planets, Jupiter. These fragments of rock, up to 100 km in diameter, form the asteroid belt. Jupiter is the largest planet in the solar system. It has a spectacular appearance, with a swirling atmosphere and its Great Red Spot. It also has sixteen moons in orbit around it.

Further out from the Sun, the planets become colder and this affects their composition. Saturn, the sixth planet, is a very bright yellow object in the sky. Its rings were first noticed by Galileo in 1610, using one of the first telescopes. These rings are made up of dust, ice and rock. When Voyager 2 flew past Saturn in 1981 it counted more than 100 000 small rings. In addition to its rings, Saturn has more than twenty satellites in orbit around it.

Uranus also has rings and a total of fifteen satellites, five of which are large moons. Uranus and the eighth planet, Neptune, are very similar in size and composition. Being further away from the Sun, Neptune takes twice as long as Uranus to complete an orbit. Voyager found six moons orbiting Neptune in addition to the two that were already known.

Neptune was the last planet to be visited by Voyager, leaving Pluto, the outermost known planet, still to be photographed at close range. Not much is known about Pluto, the most distant planet. Pluto has one known moon.

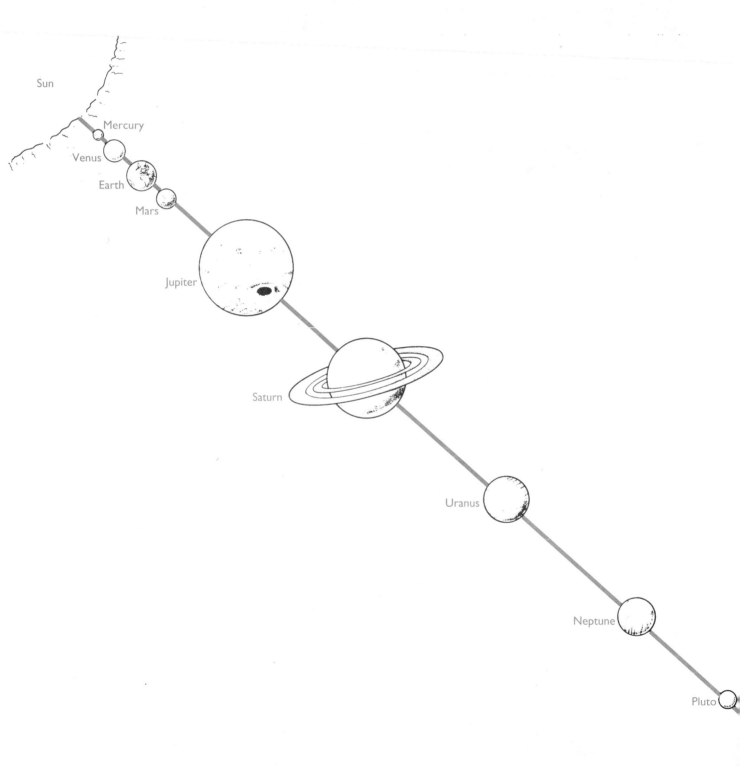

The scale of the solar system is vast and it is very difficult to fit the Sun and planets on the same diagram. The figure gives an idea of the relative sizes and distances of the planets from the Sun.

Now test yourself

The table gives information about the planets in the solar system.

Planet	Diameter Earth=1	Mass Earth=1	Surface gravitational pull Earth=1	density (g/cm^3)	Average distance from Sun Sun–Earth =1	Period of orbit (years)	Number of moons
Earth	1	1	1	5.5	1	1	1
Jupiter	11.2	318	2.6	1.3	5.2	11.9	16
Mars	0.5	0.1	0.4	4	1.5	1.9	2
Mercury	0.4	0.06	0.4	5.4	0.4	0.2	0
Neptune	3.8	17	1.2	2.3	30	165	8
Pluto	0.2	0.002	0.2	0.4	39	248	1
Saturn	9.5	95	1.1	0.7	9.5	29.5	more than 20
Uranus	3.9	15	0.9	1.6	19	84	5
Venus	0.95	0.8	0.9	5.2	0.7	0.6	0

2 (a) Whereabouts in the solar system are the densest planets found?

(b) Which is the smallest of the outer planets?

(c) The planet with the strongest gravitational pull is not the most dense. Which feature of this planet gives it the strongest gravitational pull?

(d) Why is Pluto the odd one out of the outer planets?

|N ORBIT

The Moon goes round the Earth and the Earth goes round the Sun. What does the Sun go round? The Sun is one of many stars in a spiral galaxy called the Milky Way. The Milky Way is itself spinning round and our Sun would complete an orbit once every 200 million years at its current speed and position in the galaxy. The speed of our Sun is 270 km/s, so it has a long way to go to complete one orbit.

All this rotation is due to the gravitational force that exists between any two objects. If you tie a rubber bung onto the end of a piece of string, you can whirl it round your head in a circle; but it only keeps going round you as long as you are pulling on the string. The motion of planets in the solar system is due to the gravitational pulling forces between the Sun, the planets and the moons.

The diagram, shows the relative strength of the Sun's gravitational pull at Mercury, Venus and Earth.

As you can see, the Sun's gravitational pull gets weaker at planets further away. The strength of the Sun's gravitational pull determines the speed of a planet in its orbit; the closer a planet is to the Sun, the faster it moves.

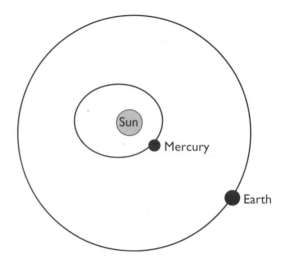

All the planets move in orbits that are ellipses, like circles that have been squashed a bit. For all the planets except Mercury and Pluto the ellipses are very near to being circles so the distance between the Earth and the Sun does not change very much during a year. Mercury's orbit takes it closer to the Sun at some times of year and this affects its speed.

As Mercury gets closer to the Sun it speeds up in its orbit, and it slows down as it gets further away. Pluto behaves in a similar way.

At Jupiter, the Sun's pull is even weaker than at the Earth, but the actual force that pulls Jupiter towards the Sun is bigger than that pulling Earth; this is because Jupiter is more massive than the Earth so there is a lot more matter for the Sun to pull on! Jupiter still moves more slowly than the Earth because the mass of a planet does not affect the speed of its orbit; its distance from the Sun is the key factor.

If the Sun's pull keeps the planets going round it, what keeps the Moon going round the Earth? The answer is the Earth's pull. Just as the Earth pulls us towards it, it also pulls the Moon.

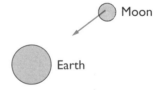

The Moon is the Earth's natural satellite, but the Earth also has hundreds of artificial satellites that do a wide variety of jobs. The first artificial satellite, Sputnik 1, was put into orbit by Russia in 1957. Russia now has a permanent manned space station, Mir, in orbit.

Satellites are used to take photographs of the Earth at regular intervals so that weather forecasters can see the changes in weather patterns. Satellites are also used for television broadcasting and for communicating by telephone and radio between places that are 'hidden' by the Earth's curvature. Ships and aircraft use satellites for navigating and some countries use satellites to spy on others. Telescopes such as the Hubble telescope have a much better vision in orbit around the Earth, where the light they receive from stars has not been distorted by passing through the Earth's atmosphere.

Much of our recent new knowledge of the solar system has come from photographs taken by spacecraft and sent back to Earth as radio signals. Spacecraft have landed on Mars and Venus and flown close enough to photograph all the other planets except for Pluto.

Now test yourself

3 The diagram shows Pluto in its orbit round the Sun.

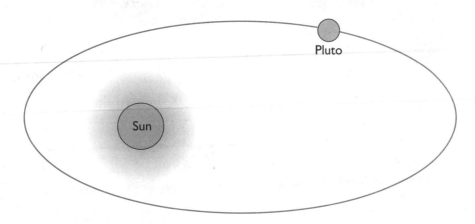

(a) Add an arrow to Pluto to show the Sun's gravitational force acting on it.

(b) What happens to the size of the force as Pluto approaches the Sun?

(c) How does this affect Pluto's speed?

4 In the Gulf war, the allied countries were able to watch the movement of Iraqi troops and military equipment from thousands of miles away. Explain how they were able to do this.

CHAPTER 14

*E*nergy resources and transfer

E NERGY FLOW

Everything we do needs energy. Eating, sleeping, walking, riding a bike, travelling on a bus, watching television; none of these things can happen without energy. In fact, without energy, nothing at all can happen. Every event, every single change that takes place in the world, involves movement. And to make things move .... energy is needed.

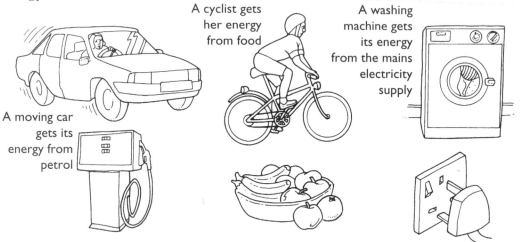

Everything that takes in energy, also gives it out. Whenever any thing is working, energy is flowing through it. The diagrams below show how most of the energy that flows out of an object at work is usually in the form of heat.

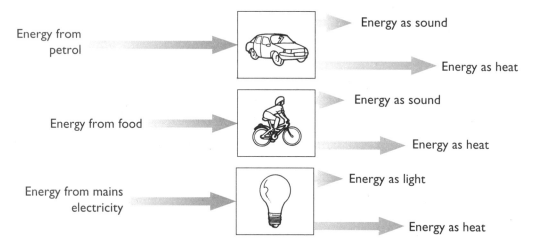

The diagrams show the energy going in and the energy going out, but each object also has some energy.

The car and cyclist both have energy because they are moving; this is called movement or kinetic energy. This energy comes from their fuel or food supply when they set off or speed up, and they have to get rid of it when they slow down or stop. In this case it again ends up as heat: in the brakes, road and tyres.

The light bulb also has some energy because it is hot; it takes this energy from the mains supply when it is switched on and it loses it again to the surroundings as it cools down after it has been switched off.

Now test yourself

1 Draw an energy flow diagram showing the energy going into and out of a television set.

WHAT HAS ENERGY?

Anything that can cause movement has energy.

❶ **Sound energy** The energy in a sound wave makes your eardrum vibrate.

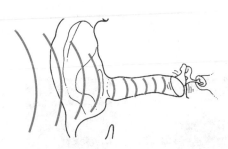

❷ **Light energy** The energy of light waves can be used to generate electricity and make a motor turn round.

❸ **Heat energy** The molecules in a glass of hot water have more energy than those in a glass of cold water; they are moving faster.

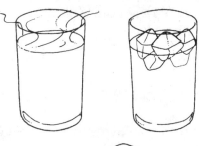

❹ **Electrical energy** Electric currents have energy; the electric current makes the motor turn.

❺ **Chemical energy** All fuels store energy; a chemical reaction with oxygen releases this energy.

❻ **Gravitational potential energy** When something is lifted up, it gains energy.

⑦ **Kinetic energy** All moving things have energy; they can cause other things to move.

⑧ **Elastic potential energy** The energy stored in a rubber band can make a toy move.

JAMES PRESCOTT JOULE (1818–1889)

The unit of energy is named after James Joule in recognition of his work on energy transfer and energy conservation. He was born in Salford and was a brewer by trade.

When he was 19 he built an electric motor. He was trying to find an alternative to the steam engine to power machines, but he was not very successful in doing this. Later he built a dynamo and using this he discovered his law of electrical heating.

Joule is also famous for discovering that when a gas expands it cools. All refrigerators and air conditioners use this principle.

Joule was a very enthusiastic scientist. Even on his honeymoon he tried to measure the temperature at the top and bottom of a waterfall to find out if the energy transfer as the water fell caused it to get warmer.

Now test yourself

2 Name the type or types of energy that the following things have:

(a) a firework rocket before it has been lit

(b) the space shuttle in orbit around the Earth

(c) a train travelling at 125 mph

(d) a bath full of hot water.

E NERGY STORAGE AND TRANSFER

Fuels such as coal, oil and gas are long-term stores of energy. When they burn, this energy is transferred into heat and light. The energy in these fuels is in a very concentrated form; a lot of energy is stored in a small amount of space.

We can also store energy as chemical energy in batteries. The energy needed to start a car engine is stored in its battery; this energy is transferred into kinetic energy of the engine when it is started. The energy is much less concentrated than that in a fuel such as coal. Battery-powered vehicles such as milk floats need large batteries to store enough energy to be able to travel a few miles.

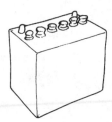

Electricity cannot be stored, so electricity generating companies store their surplus energy as **gravitational potential energy** . At night, when there is not much demand for electricity, the surplus electrical energy is transferred to gravitational potential energy. This is done by pumping water from a low reservoir into a high one. The gravitational potential energy can be transferred back into electrical energy during the daytime when there is a lot of demand for electricity.

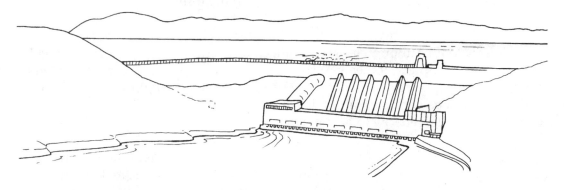

Although it cannot be stored, electricity is so widely used because it provides energy which can easily be transferred to useful forms such as heat, light, sound and movement that we need for our everyday comfort. Electrical devices that we use at home and at work are designed to transfer the energy from an electric current into these useful forms.

Now test yourself

3 (a) Here is a list of devices that take in energy from the mains supply. Write down the form or forms of energy that each one is **designed** to give out:
(i) a radio
(ii) a lamp
(iii) a hairdryer
(iv) a television set
(v) a kettle.

(b) What energy transfer is going on here?

E NERGY CONSERVATION

Some devices are very good at transferring the energy into the form we want. Others are not so good. The diagrams show the energy flow through a lamp and a kettle element in one second.

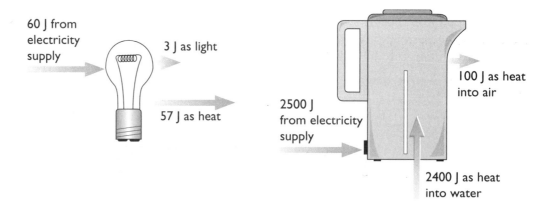

Lamps and kettles, like other things that transfer energy, do not use any energy up. The same amount of energy flows out of them as flows into them. This is known as 'conservation of energy'. The diagrams also show that the kettle wastes a lot less energy than the lamp. The lamp filament has to be very hot before it gives out light, so a lot of heat is produced from the energy that is supplied to it. Fluorescent lights do not get so hot and so they waste less energy in the form of heat.

Even the 3 J of light energy produced by the lamp each second does not exist as light for very long. Although it may be reflected several times, most of it is quickly absorbed by the surfaces it meets. When you turn the light out in a room, it soon becomes dark! The surfaces that absorb this energy become slightly warmer. This warming would be very difficult to detect with an ordinary thermometer; lights do not give out enough energy to heat a room appreciably.

Sound is another form of energy that does not last for very long. When a sound has been made, it too has its energy absorbed by the surfaces it meets, making them slightly warmer. However, you would have a difficult job persuading your parents that you need to play the hi-fi loud just to warm the room up!

Almost all the energy that we transfer each day, both from our bodies and the things that we use, ends up as heat energy in our surroundings; in the buildings we inhabit, the air, and the outdoors. We cannot get this energy back very easily; it is much easier to use more energy from a fuel or the electricity supply to heat our homes than to try to extract the energy from the air and the ground outside.

Now test yourself

4 Complete the energy flow diagram for a radio.

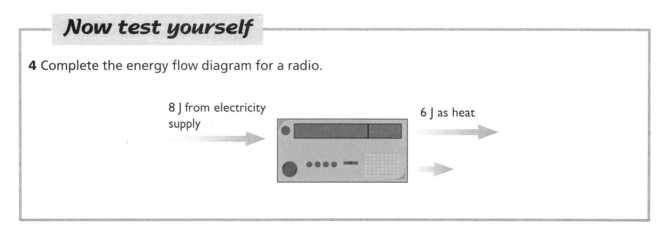

E NERGY RESOURCES

Without the Sun, there would never have been any life on the Earth. We depend on energy from the Sun to keep on living. It gives us warmth and light and also provides the energy for plants to grow.

We need plants for our food. Plants are a **renewable** energy source – we can grow them over and over again.

Some energy sources are **non-renewable** because they took so long to form that no more can be made within our lifetime. All fossil fuels, coal, gas and oil, are non-renewable. The diagram illustrates some of the Earth's energy resources.

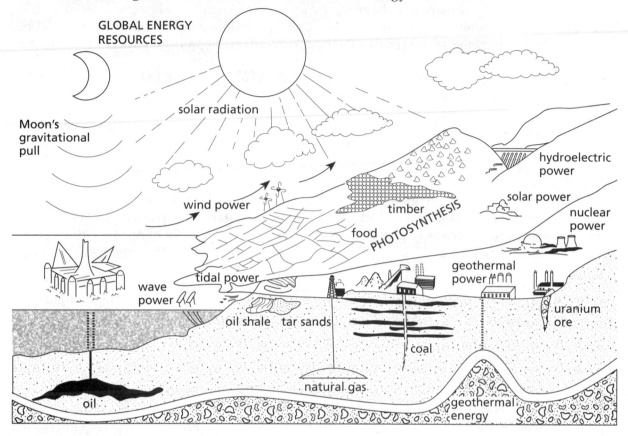

The energy in coal has been stored for millions of years. Energy from the Sun was transferred into the chemical energy of giant fern-like plants. The fossils of these ferns can often be seen in lumps of coal. The decaying remains of these plants formed layers that became covered in mud, sand and clay. Over millions of years the pressure from on top, combined with heat from the Earth below, caused coal to form from the remains of the ferns.

Oil and natural gas are also fossil fuels. They formed in a similar way to coal from the remains of animals that lived in the sea. When the animals died, their bodies formed layers on the seabed that were subjected to pressure from above and heat from below. Like coal, the energy stored in gas and oil came from the Sun and was stored in the plants that the animals ate.

The fossil fuels that took millions of years to form are being used up very rapidly. In Britain there are large known reserves of coal that could last for two hundred years, but a lot of it will be very expensive to mine because it is in narrow seams that are not easy to get to.

The known reserves of gas and oil will not last as long as those of coal, but new reserves are being found every year. To make these reserves last, we need to use them as efficiently as possible and to find alternative energy sources for transport and for generating electricity.

We have to make more use of the energy from the Sun that reaches us each day. One way to do this is to grow more plants as fuels. The energy stored in biological materials, or **biomass**, can be used in several ways. Wood can be grown quickly using fast-growing trees. Wood produces a lot of smoke when it burns, but if the tar is removed first by heating the wood without an oxygen supply, it makes charcoal which is a very clean fuel.

Countries such as Brazil that do not have natural reserves of oil use ethanol as a fuel for vehicles. Ethanol can be made by fermenting sugar. In a hot country like Brazil large amounts of sugar can be grown each year to make a new supply of ethanol. Both wood and ethanol are **renewable** because more can be produced in a short time-scale.

Energy from the Sun also makes the wind blow. The energy in moving wind has been used for thousands of years to turn windmills. Recently in Britain there has been a large increase in the number of wind turbines that generate electricity. These are sometimes built in groups called wind farms. In some exposed areas such as the Yorkshire moors the wind rarely stops blowing so wind farms can be relied on to give a constant supply of electricity.

Some people argue that wind farms are environmentally friendly because they do not burn any fuel and do not produce any waste gases. People who live near wind farms sometimes say that they are ugly and they make a lot of noise – a different kind of pollution.

The sea provides two important energy resources: waves and tides. Waves are caused by the wind, and tides are due mainly to the effect of the Moon pulling on the Earth although the Sun has a small effect as well. Attempts to extract energy from waves have been abandoned because of the very high cost of making the equipment to do this.

Like energy from the wind, energy from tides has been used for hundreds of years to drive mills. It has the advantage of being predictable and reliable as it depends only on the Moon orbiting the Earth, something which is not likely to stop happening in the foreseeable future!

In countries with more sunshine than Britain, hot-water tanks on the roofs of houses are a common sight. Energy from the Sun is used to heat water as it passes through some copper pipes. The pipes are black to absorb as much of the Sun's radiant energy as possible.

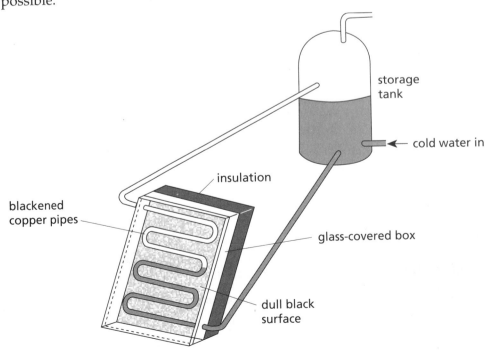

storage tank

cold water in

insulation

blackened copper pipes

glass-covered box

dull black surface

Generating Electricity

Since the first power station opened in Britain over a hundred years ago, electricity has become an energy source that we rely on more and more. Batteries provide a useful portable energy source for, say, a Walkman, but they are expensive and they use up valuable chemicals.

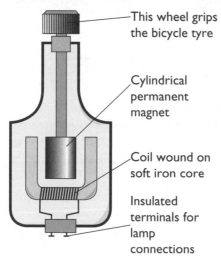

This wheel grips the bicycle tyre

Cylindrical permanent magnet

Coil wound on soft iron core

Insulated terminals for lamp connections

Whatever the energy source, mains electricity is generated by making a magnet rotate inside coils of wire. A bicycle dynamo uses a permanent magnet to generate the electricity needed to light two lamps. Power stations use an electromagnet that is driven round at 50 revolutions each second.

The electromagnet in a generator is driven by a turbine. A turbine is a set of blades that is driven by a moving fluid. The wind turbine shown in the photograph has just two blades that are driven round by the wind.

Most of our electricity is generated using steam-driven turbines. The energy from a nuclear reactor or a coal-fired boiler is used to change water into high-pressure steam. Energy from the steam keeps the turbine blades rotating. Some of the newer power stations burn natural gas. Instead of using the gas to change water into steam, the hot gases produced when the gas burns are used to turn the turbine blades.

Water-driven turbines are used in hydroelectric power stations. The gravitational potential energy of water is transferred to kinetic energy as it falls through pipes leading from a lake or reservoir to the turbines. Kinetic energy from the fast-moving water is used to generate electricity as it passes through the turbines, so it emerges as slow-moving water that is returned to a river or lake.

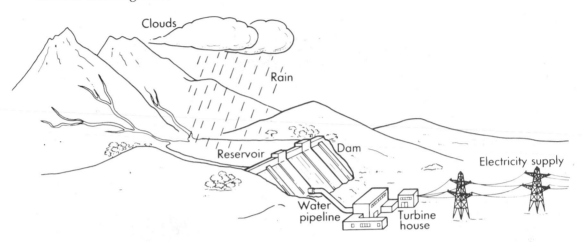

Hydroelectric power stations provide a large proportion of the electricity generated in Scotland, where there are a large number of high-up lakes and reservoirs that can be used. Hydroelectric power uses the Earth's natural water cycle that is driven by energy from the Sun.

Electricity from the tides is a renewable energy source that should be everlasting, but the cost of building a suitable power station is very high and the effect on the environment could be considerable.

If water turbines were built into a dam across the River Severn estuary they could supply as much electricity as all our nuclear power stations do together. A similar scheme has been operating on the La Rance estuary in France for thirty years. The diagrams show how electricity is generated by both the incoming and outgoing tides.

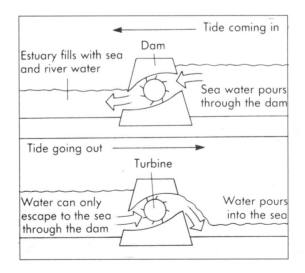

Far fewer power stations would be needed if we could generate more electricity directly from the Sun. Cells that use the energy from light to produce electricity are common in calculators and they provide the energy for satellites, but they would be very expensive to produce on a scale large enough to provide energy for the mains supply. They also take up a large area.

A new type of solar cell has recently been developed which promises to be much cheaper to manufacture. It only transfers 10 per cent of the energy from light into electricity, but it is transparent so it can be built into windows. Houses in the future could have windows that generate electricity as well as letting in light.

After its launch from the space shuttle, this satellite will keep its batteries charged by using energy from the Sun to generate electricity in the solar panel.

Now test yourself

5 Refer to the energy resources diagram on page 141. Divide the energy resources into two lists: renewable and non-renewable.

6 Some farms are self-sufficient in energy. The manure from animals is placed in a tank. Bacteria feed on this manure and produce methane gas which is burned to generate electricity.

(a) What is the original source of this energy?

(b) Is this energy resource renewable or non-renewable? Explain your answer.

(c) The burning of methane gas transfers energy. Describe the energy transfer that takes place.

(d) In the generating process, 40 per cent of the available energy is transferred to electricity. Describe what happens to the energy that is not transferred to electricity.

LEVEL DESCRIPTIONS
*P*hysical Processes

At the start of Key Stage 3 the majority of pupils will have reached at least Level 4 in Science. By the end of Key Stage 3 most pupils should be within the range of Levels 4–7. Levels 5–6 are the target for 14-year-olds. Level 8 is the standard reached by very able pupils.

Use our checklist to assess the Level reached, by ticking the skills that have been mastered.

Level 4

☐ Demonstrate knowledge and understanding of aspects of physical processes drawn from the Key Stage 2 or Key Stage 3 programme of study.

☐ Describe and explain physical phenomena, such as how a particular device in an electrical circuit may be switched on or off, or how the apparent position of the Sun changes over the course of a day.

☐ Make generalizations about physical phenomena, such as motion being affected by forces, including gravitational attraction, magnetic attraction and friction, or sounds being heard through a variety of materials.

☐ Use the idea that light travels, to explain phenomena such as the formation of shadows.

Level 5

☐ Demonstrate an increasing knowledge and understanding of aspects of physical processes drawn from the Key Stage 2 or Key Stage 3 programme of study.

☐ Begin to apply ideas about physical processes to suggest a variety of ways to make changes, such as altering the current in a circuit or altering the pitch or loudness of a sound.

☐ Begin to use some abstract ideas in descriptions, such as forces being balanced when an object is stationary, or objects being seen when light from them enters the eye.

☐ Use models to explain effects that are caused by the movement of the Earth, such as the length of a day or year.

Level 6

☐ Demonstrate understanding of how to apply some abstract ideas about physical processes drawn from the Key Stage 3 programme of study.

☐ Use abstract ideas in descriptions and explanations, such as electric current being a flow of charge, the sum of several forces determining changes in the direction or the speed of movement of an object, or wind and waves being energy resources available for use.

☐ Recognize that many physical concepts, such as the transfer of energy by light, sound or electricity, or the refraction and dispersion of light, apply in a variety of contexts, and describe some of these.

☐ Recognize that a number of factors may have to be considered when phenomena, such as the relative brightness of planets and stars, are explained.

Level 7

☐ Use knowledge and understanding of physical processes drawn from the Key Stage 3 programme of study, to make links between different phenomena.

☐ Make connections between electricity and magnetism to explain phenomena such as the strength of electromagnets.

☐ Use some quantitative definitions, such as those for speed or pressure, and perform calculations involving physical quantities, using the correct units.

☐ Apply abstract ideas in explanations of a range of physical phenomena, such as the appearance of objects in different colours of light, the relationship between the frequency of vibration and the pitch of a sound, the role of gravitational attraction in determining the motion of bodies in the solar system, or the dissipation of energy during energy transfers.

Level 8

☐ Demonstrate an extensive knowledge and understanding of the physical processes in the Key Stage 3 programme of study.

☐ Use models to describe and explain phenomena, such as the flow of charge in parallel circuits or the passage of sound waves through a medium.

☐ Use quantitative relationships between physical quantities in calculations that many involve more than one step.

☐ Offer detailed and sometimes quantitative interpretations of graphs, such as speed–time graphs.

☐ Consider ways of obtaining data, such as data relating to the solar system, and use knowledge of physical processes to explain patterns.

☐ Consider physical phenomena from different perspectives, such as relating the dissipation of energy during energy transfer to the need to conserve limited energy resources.

Exceptional performance

☐ Demonstrate both breadth and depth of knowledge and understanding of the physical processes in the Key Stage 3 programme of study and draw on aspects of the Key Stage 4 programme of study when describing and explaining physical phenomena.

☐ Make effective use of a range of quantitative relationships between physical quantities such as resistance, acceleration or wave speed.

☐ Understand how models, such as the wave model, are useful in explaining physical phenomena, such as the transmission of light and sound.

☐ Apply understanding of physical phenomena to a wide range of systems, such as recognizing the role of gravitational attraction in determining the movement of satellites, planets and stars.

☐ Recognize the importance of quantitative data and make effective use of this when considering questions such as energy efficiency.

*P*ractice National Test Questions

These are examples of the kind of questions you will meet in your Key Stage 3 National Test papers. Practising these example questions will help you prepare for your test and to do as well as you can. So that you can check how well you are doing, the approximate National Curriculum level for each question or part of question is given in brackets.

L IFE PROCESSES AND LIVING THINGS

1 (a) The diagram shows a root hair cell from a plant.
(i) Use words from the following list to label the diagram:
cell membrane   cell wall   nucleus   [LEVEL 6]

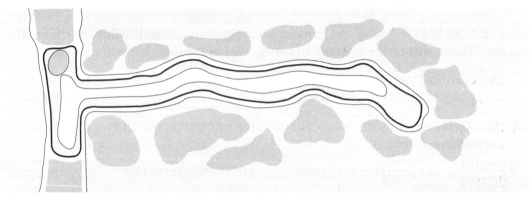

(ii) How does the structure of the cell help the cell to absorb water from the soil?
[LEVEL 7]

(b) The diagram on the left shows a cell from the leaf of a plant.
The cell is damaged by freezing. Suggest a reason why the cell is damaged.
[LEVEL 6]

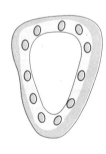

2 The graph below shows how the number of cigarettes smoked changes the risk of death of a person from lung cancer.

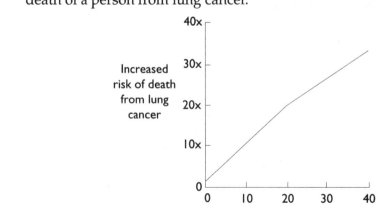

(a) What effect does smoking more cigarettes have on the risk of getting lung cancer?

(b) What is the increased risk of dying from lung cancer if a person smokes 10 cigarettes a day?   [LEVEL 5]

3 In this question the organs in the following list can be used to answer the questions which follow:
brain   heart   liver   lungs   stomach

(a) The diagram shows that the space inside a human body is divided into two parts – the abdomen and the thorax.

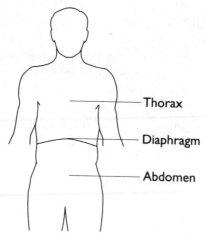

(i) Write down the names of the organs in the list which are in the abdomen.

(ii) Write down the names of the organs in the list which are in the thorax.

[LEVEL 4]

(b) Which of the organs in the list:
 (i) pumps blood around the body?
 (ii) exchanges gases between the blood and air?
 (iii) digests the food?

[LEVEL 5]

4 This question is about the parts of a flowering plant.
 Write down the job of each of the following parts:
 (a) leaf
 (b) root
 (c) sepal
 (d) anther.

[LEVEL 5]

5 The local council produces new plants to use for planting out flower beds on roundabouts etc. At the end of the summer they store geranium plants over the winter. Next spring they produce new geranium plants. They can do this in two ways:
 1 taking stem cuttings of the stored plants
 2 from seed purchased from a seed firm.
 Give two advantages of taking stem cuttings.

[LEVEL 7]

6 The diagram shows a tomato plant.

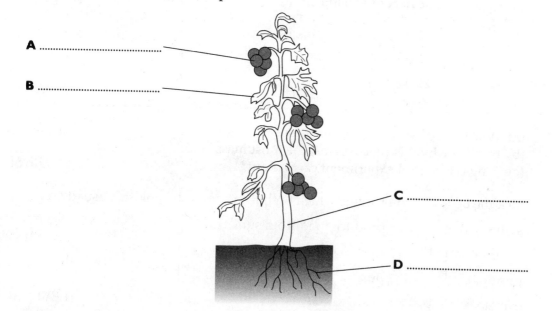

A ....

B ....

C ....

D ....

149

(a) Use the **best** words from this list to label the diagram:
    flower   fruit   leaf   root   stem   [LEVEL 4]
(b) Which part of the plant
    (i) has seeds inside?
    (ii) takes in water?   [LEVEL 5]
(c) A gardener growing tomatoes in a greenhouse uses a paintbrush in the middle
    of the day. He pushes the bristles of the brush into a number of the flowers.
    (i) Why does he push the bristles of the paintbrush into the flowers?
    (ii) Why is it an advantage to do this?
    (iii) Why is the middle of the day most suitable?   [LEVEL 6]

7 The diagram shows test tubes set up to find out how plants lose water.

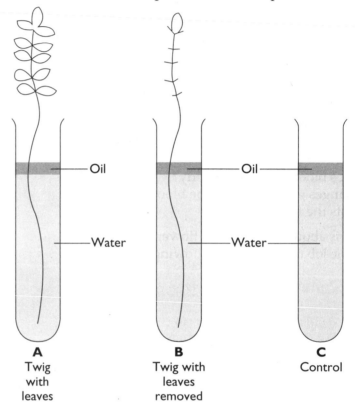

The test tubes and contents were weighed at the start of the experiment and
again the next day. The results are shown in the table.

Tube	Mass of tube and contents at start of experiment (g)	Mass of tube and contents after 1 day (g)
A	32.0	28.5
B	27.0	26.5
C	25.0	25.0

(a) Which tube has lost the most mass?   [LEVEL 4]
(b) Explain why this tube has lost the most mass.
(c) Why is a control experiment carried out?   [LEVEL 5]

8 Tick the boxes by **four** things which both the cat and the goldfish usually do.

swim	☐	produce food from sunlight	☐
breathe	☐	feed	☐
grow	☐	purr	☐
excrete	☐		

[LEVEL 3]

9 The following key can be used to identify some seaweeds found on a beach.

1	Seaweed has branched fronds (leaves)	go to 3
	Seaweed does not have branched fronds	go to 2
2	Frond has fingers	Oarweed
	Frond does not have fingers	Tangleweed
3	Fronds have bladders (air bags)	go to 4
	Fronds do not have bladders	Saw wrack
4	Fronds divide evenly into two	Ladderwrack
	Fronds do not divide evenly into two	Knotted wrack

Use this key to identify the five seaweeds below.

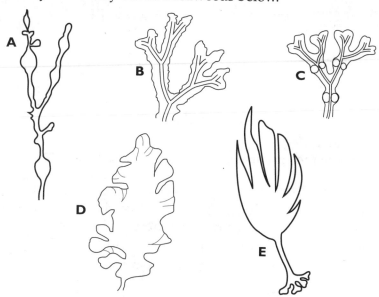

[LEVEL 4]

10 Fertilization in mammals happens when the nucleus of a sperm cell joins with the nucleus of an egg cell.
   (a) In which organs are the following produced?
      (i) Sperm cells.
      (ii) Egg cells. [LEVEL 5]
   (b) The fertilized egg then divides into many cells before developing into an embryo. Where in the body of a mammal does the embryo develop?

Joe and John are identical twins. It is almost impossible to tell them apart.
   (c) Use your knowledge of genetics to explain why they are so alike.
   (d) Twins need not be identical nor even be the same sex. Explain how this is possible.
   (e) If identical twins are separated at birth and brought up in different families, the twins can have different masses (weights). Explain why this may be the case.

[LEVEL 7]

11 The diagrams show two different species of grasshopper.
   (a) Name the genus to which both these grasshoppers belong.
   (b) From the diagrams, give a characteristic which shows that grasshoppers are insects.

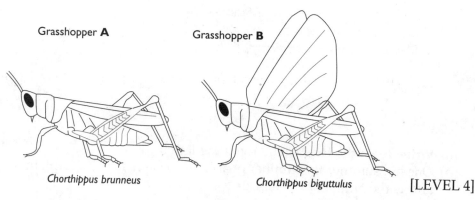

Grasshopper **A**   Grasshopper **B**

Chorthippus brunneus   *Chorthippus biguttulus*   [LEVEL 4]

12 In an oak wood there are bluebells which flower in early spring. Woodpeckers . feed on insects under the bark of the trees.

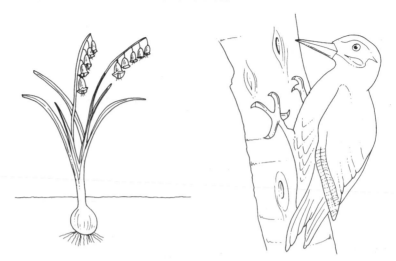

Write down **two** advantages of the bluebell and **two** advantages of the wood-pecker which enable the organisms to survive in their environments.

[LEVEL 6]

13 Trilobites are extinct animals. When they were alive, they had scales on their backs for protection. These scales have been preserved as fossils. The diagrams show a typical trilobite and a cross-section through the trilobite when it was alive.

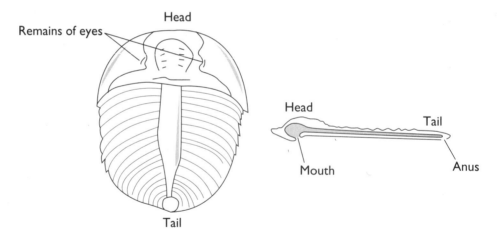

(a) Describe **three** ways in which trilobites were adapted to live on the sea bottom.

[LEVEL 5]

(b) Explain how fossils such as trilobites help geologists to estimate the age of rocks.

[LEVEL 8]

14 The diagram shows a food web.

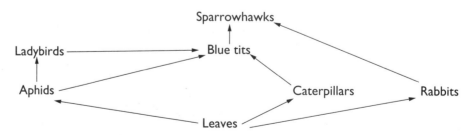

(a) Write down the food chain involving ladybirds. [LEVEL 4]

(b) One hot summer the number of aphids inceases rapidly. Explain how this affects the number of ladybirds.

(c) A large number of rabbits are killed by disease. Explain why this has little effect on the numbers of sparrowhawks.

(d) Explain why the death of a large number of rabbits causes the ladybird population to increase.

[LEVEL 7]

15 A pigeon eats slugs. The pigeon is called the **predator** and the slug is called the **prey**. The figure shows eight organisms.

Not to scale

Complete the following to show other predator–prey relationships. One has been done for you.

Prey	Predator
fly	spider

[LEVEL 4]

MATERIALS AND THEIR PROPERTIES

1 The diagrams show different materials inside four different beakers.

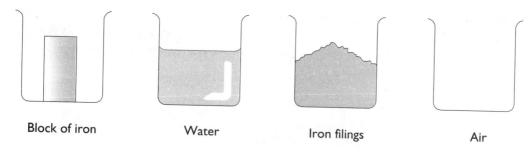

Block of iron   Water   Iron filings   Air

The table shows some of the properties of these four materials.

(a) Which property does only air have?

Property	Block of iron	Water	Iron filings	Air
Can you put your fingers through it?	no	yes	yes	yes
Can it be poured?	no	yes	yes	yes
Does it have a fixed shape?	yes	no	no	no
Can it be squashed?	no	no	no	yes

(b) Which material has a fixed shape?

(c) Complete the table by filling in the state of the material.

Material	Solid, liquid or gas
block of iron	
water	
air	

(d) Explain why it is possible to pour the iron filings but not the iron block.

[LEVEL 4]

2   The diagram below shows apparatus set up to make pure water from blue ink.

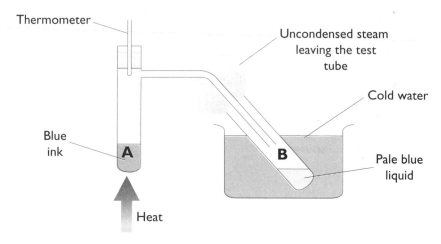

(a) What changes of state are taking place at (i) **A** and (ii) **B**?
(b) Circle the highest temperature shown on the thermometer during the experiment.
    0°C   20°C   50°C   100°C   150°C
(c) Give two reasons why the diagram shows that the method is not very efficient.

[LEVEL 5]

3 Chromatography can be used to separate the metals present in an alloy (mixture of metals). The alloy called German silver is reacted with nitric acid and the resulting solution of metal nitrates used for the experiment. The diagram shows the chromatogram.

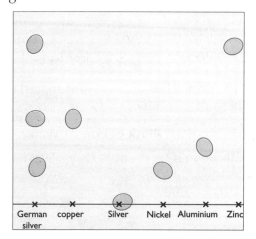

(a) Give the names of three metals in German silver.
(b) Give the names of two metals **not** in German silver.

[LEVEL 5]

4 The table shows information about three substances.

Substance	Appearance	Change on heating in air	Formed when substance burned	Acidity/alkalinity of product
A	yellow solid	burns easily	a gas	acid
B	grey solid	burns easily	a gas and a liquid	neutral
C	grey solid	burns easily	a solid	alkali

(a) Which **two** substances could be elements? Explain your answers.

(b) Why is substance **A** likely to be a non-metal?

[LEVEL 7]

5 The table shows some of the properties of solids, liquids and gases.

Property	Solid	Liquid	Gas
hard	yes	no	no
can flow	no	yes	yes
can be squashed	no	no	yes

Use the table to finish the following sentences. The first one has been done for you.
Brick is made of a *solid* because *brick is hard and cannot be squashed easily*.

(a) The fluid in a hydraulic brake system of a car is a ....

because ....

(b) A bicycle tyre is filled with a ....

because ....

[LEVEL 4]

6 Kim made observations of two pieces of rock. These are shown in the table.

	Rock A	Rock B
Colour	Some white crystals and some black ones	White crystals with brown veins through the rock
Size of crystals	Large	Small

(a) What evidence is there in her table to suggest that Rock A is made of more than one mineral?

(b) What evidence is there in her table to suggest that Rock A cooled slowly?

(c) Rock B is marble. Marble is a metamorphic rock formed from the sedimentary rock limestone.

What happened to turn limestone into marble?

[LEVEL 6]

7 The key below can be used to identify minerals.

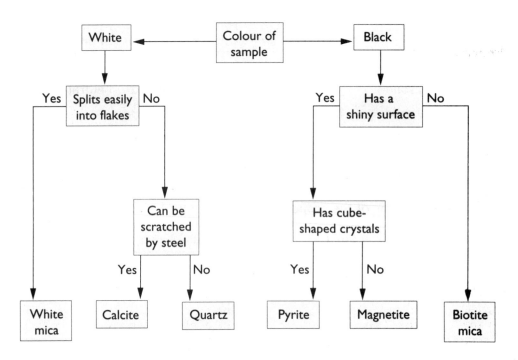

Use the key to identify minerals **C** and **D**.
C is white, does not split easily into flakes and is harder than steel.
D is black, has a shiny surface and cube-shaped crystals. [LEVEL 4]

8 The figure shows how a pillar of rock which stands in the desert has changed over a thousand years.

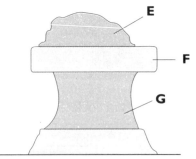

(a) What might have caused the pillar to change shape?
(b) Which layer of rock, **E**, **F** or **G**, is likely to be the hardest? Explain your answer. [LEVEL 6]

9 Rocks are made up from minerals. The table below compares properties of five minerals: calcite; fluorite; barite; galena and sphalerite.

Property	Mineral				
	calcite	fluorite	barite	galena	sphalerite
colour	white, pink, colourless	blue, yellow, green, colourless	white, pink	dark grey	black to brown
streak	white	white	white	thick grey	pale brown or off-white
lustre	glassy	glassy	glassy to dull	metallic	metallic to glassy
hardness	3	4	3.5	2.5	3.5
density	moderate	moderate	surprisingly dense for a white material	very dense	moderate
acid test	fizzes violently	nil	nil	**do not test**	nil
habit (shape)	'dog-tooth', shape, or more often rhombus-shaped cleaved fragments	cubic sometimes with corners missing	often 'cocks-comb' structure, otherwise layered	often broken into cubic cleavage fragments	good shapes rare

Use the table to answer the following questions:
(a) Which mineral should not be tested with acid?
(b) Which test would you use to tell apart the minerals in each of the following pairs?
  (i) calcite and fluorite
  (ii) fluorite and barite
  (iii) galena and sphalerite.
(c) How would you separate a mixture of galena and fluorite?   [LEVEL 6]

10 The table shows results obtained when zinc oxide and zinc carbonate are heated.

	Appearance before heating	Appearance after heating	Mass before heating (g)	Mass after heating (g)
zinc oxide	white powder	turns yellow but turns white again on cooling	2.3 g	2.3 g
zinc carbonate	white powder	white powder	2.3 g	2.0 g

Why does the information in the table suggest zinc carbonate is decomposed on heating but zinc oxide is not?   [LEVEL 6]

11 The table gives the products of electrolysis of aqueous solutions of sodium chloride and copper(II) nitrate.

	Product at negative electrode	Product at positive electrode
sodium chloride solution	hydrogen	chlorine
copper(II) nitrate solution	copper	oxygen

Finish the following table by predicting what you would **see** at each electrode when electrolysis of sodium nitrate solution and copper(II) chloride solution takes place.

	Seen at negative electrode	Seen at positive electrode
sodium nitrate solution		
copper(II) nitrate solution		bubbles of greenish gas

[LEVEL 6]

12 Sara tested four metals with different metal salt solutions.
She put a tick (✔) if there was a reaction and a cross (✗) if there was no change.
Her results are shown in the table.

	Magnesium chloride solution	Copper chloride solution	Iron chloride solution	Zinc chloride solution
magnesium	✗	✔		
copper	✗	✗		
iron	✗	✔		
zinc	✗	✔		

(Refer to the reactivity series on page 81.)
(a) Finish the table.
(b) Zinc reacts with copper nitrate solution. Write a word equation for this reaction.
(c) When Sara tested another metal lead with the same four metal salt solutions, it only reacted with copper chloride solution.
   (i) What would be seen when it reacted?
   (ii) What does this experiment tell her about the position of this metal in the reactivity series? [LEVEL 7]

13 Here is a list of words to describe types of chemical reaction:
endothermic neutralization oxidation polymerization reduction
Choose the best word in the list for each of the following reactions:
(a) The reaction of an acid with an alkali.
(b) Making poly(ethene) from ethene.
(c) Making iron from iron oxide in a blast furnace.
(d) Burning hydrogen gas in oxygen. [LEVEL 8]

14 Gill placed samples of four metals **P, Q, R** and **S** in turn into aqueous solutions of the nitrates of the other three.

Here are her results.

P displaced **S**

Q displaced **P**

Q displaced **S**

R displaced **Q**

Arrange these four metals in order of decreasing reactivity. Explain why you put the metals in this order.

[LEVEL 7]

15 This question is about the preparation of zinc sulphate crystals.

Read the passage carefully and then answer the questions.

Zinc carbonate powder was added to a colourless liquid (**A**) until no more bubbles of gas were produced when fresh zinc carbonate was added. The resulting zinc sulphate solution, after removing any excess zinc carbonate, was concentrated. This solution was left for a few days and crystals of zinc sulphate were formed.

(a) Name

(i) the gas formed when zinc carbonate is added

(ii) the colourless liquid **A**.

(b) Why were no more bubbles produced when extra zinc carbonate was added to **A**?

(c) How could the excess zinc carbonate be removed from the solution?

(d) The crystals formed were quite large. How could you produce small crystals?

(e) When zinc sulphate crystals are heated, steam is produced. Underline the best word to describe this change:

combustion   decomposition   evaporation   melting   precipitation

[LEVEL 8]

P HYSICAL PROCESSES

1 Write down the readings on the ammeters numbered A1 to A4.

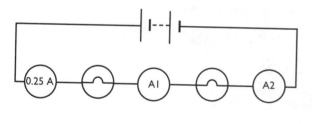 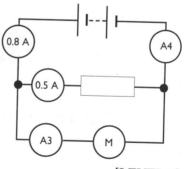

[LEVEL 5]

2 The circuit diagram shows a cell being used to light two lamps.

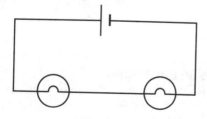

(a) Write down **two** changes that you could make to the circuit that would **increase** the current.

(b) Describe how you could adapt the circuit so that you could change the brightness of the lamps.

(c) The next diagram shows how a switch has been added to the circuit.

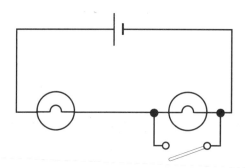

Describe what happens to **each** lamp when the switch is closed.

(d) Explain why this happens. [LEVEL 5]

3 (a) An electromagnet is made by winding a coil of wire on a metal core. Which metal is the most suitable for the core? Choose from the list below:
   brass   copper   iron   magnesium   steel

(b) Explain why this metal is the most suitable. [LEVEL 7]

4 When a firework 'rocket' is lit the burning fuel produces an upward force on the rocket.

(a) If the weight of the rocket is bigger than this upward force, what happens to the rocket?

(b) What needs to happen for the rocket to move upwards?

(c) The rocket is moving upwards at a steady speed. What does this tell you about the upward and downward forces on the rocket? [LEVEL 5]

(d) Explain what happens to the rocket when the fuel has stopped burning. [LEVEL 6]

5 The diagram shows a cycle brake. When the cyclist brakes, the rubber block is pressed against the rim of the wheel.

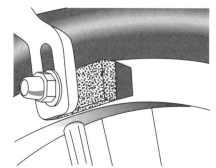

(a) What causes the wheel to slow down when the cyclist brakes? [LEVEL 4]

(b) Explain why the brakes do not work as well in wet weather.

(c) Some cycle wheel rims are made with a dimpled pattern so they are not smooth. What advantage do these wheel rims have when braking? [LEVEL 5]

6 A skier who weighs 800 N is stood on two skis which **each** have an area of 0.4 m^2.

(a) Calculate the pressure on the ground.

(b) What happens to the pressure when the skier lifts one leg (and ski) off the ground? [LEVEL 7]

7 Ice skaters cause enough pressure to melt the ice. A slim ice skater weighs only 500 N. The area of the blades on her skates is 2 cm^2. Calculate the pressure on the ice. [LEVEL 7]

8 The diagram shows a force, **F**, being used to open a hinged trap-door. The weight of the trap-door is 150 N and this force acts from the centre of the trap-door.

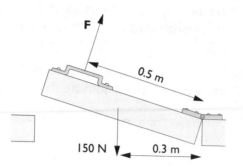

(a) Calculate the moment of the weight of the trap-door.
(b) Calculate the value of the force **F** required to make the trap-door open.

[LEVEL 7]

9 A car driver can see the car behind her by looking in a mirror.

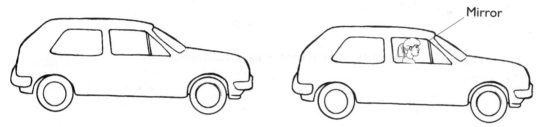

Draw on the diagram to show how the mirror enables her to see the car behind.

[LEVEL 5]

10 During a storm, a flash of lightning can be seen several seconds before the sound is heard. Why is the light detected before the sound?   [LEVEL 5]

11 (a) When white light hits a blue jumper all colours of light are absorbed except blue.
Which two **primary** colours are reflected when white light hits a yellow jumper?

[LEVEL 6]

(b) What colour would each jumper seem to be when viewed in green light?

[LEVEL 7]

12 A loudspeaker cone vibrates to produce a sound. When the same pitch sound is played louder, what changes take place to the **amplitude** and **frequency** of the vibration?

[LEVEL 7]

13 The diagram shows the Earth and Mars in orbit around the Sun.

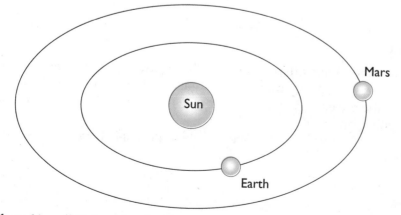

(a) Mark an X on the drawing to show where the Earth will be six months later.

[LEVEL 5]

(b) Underline the time it takes for the Earth to spin once on its axis:
one hour   one day   one month   one year

[LEVEL 4]

(c) Explain how it is possible for a person on the Earth to see Mars.   [LEVEL 5]

(d) The planets are orbiting in an anticlockwise direction in the diagram. Describe what will happen to the distance between the Earth and Mars as they continue their orbits. Explain why this will happen.   [LEVEL 6]

(e) Label the approximate positions of Mercury and Venus on the diagram.   [LEVEL 4]

14 The diagram shows light from the Sun going towards the Earth and the Moon. The line is the Earth's axis, the line joining the North and South poles.

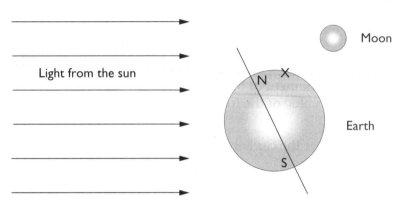

(a) Shade in the part of the Earth that is in darkness.   [LEVEL 4]

(b) Explain how a person at X is able to see the Moon.   [LEVEL 5]

(c) Explain how you can tell that it is summer in the northern hemisphere.

(d) How many hours of daylight will the South pole have during the next day?   [LEVEL 6]

15 Read the list of energy sources:

coal   tides
oil   gas
wind   charcoal

(a) Which of these could run out in the future?

(b) Five of the sources in the list obtain or obtained their energy from the Sun. Which is the odd one out?

(c) Name the two sources in the list that have energy due to movement.   [LEVEL 5]

16 In a modern gas-fired power station, 50 per cent of the energy in the gas is transferred to electric current. What happens to the rest of the energy?   [LEVEL 7]

17 Write down the name of a device that can transfer the energy from electricity into:

(a) sound

(b) heat

(c) movement

(d) light.   [LEVEL 6]

18 On a ski slope, skiers are pulled uphill by a moving rope. They then ski downhill.

(a) What type of energy do the skiers gain as they move uphill?

(b) Describe the energy transfer that takes place as the skiers set off and speed up going downhill.   [LEVEL 6]

(c) What has happened to the energy when the skiers have stopped at the bottom of the hill?   [LEVEL 7]

Answers

Chapter Three

1 Boiled potato or jacket potato (providing not fat added after).
Other forms of potato have fat added.

2 (a) (i) Cheese, roast chicken, pork chop.
(ii) White bread, brown bread, banana, fruit cake.
(b) Have similar protein contents but considerably less fat.

3 (a) Providing a vegetarian eats a full range of other foods they can intake all of
the essential vitamins and minerals from other sources. No vitamin or
mineral is only in meat.
(b) Calcium is necessary for building up teeth, bones and the skeleton. Good
sources of calcium include milk, butter and cheese.

4 (a) Brand X. Much lower fat and carbohydrate
(b) Sugar
(c) Chilled cabinet
(d) Protein 5.0 g; Fat 3.125 g; Carbohydrate 23.0 g

5 (a) Number and nature of predators
(b) Generally the more eggs produced the lower the level of care.

Chapter Four

1 Sepals 6; petals 6; stamens 9; carpels 1.

2 **Coconut** Fruit floats away from parent plant. For this reason a coconut tree can
grow on a small desert island where there are no other coconut trees.
Burdock Fruit has small hooks on the outside which fix onto animals and so are
taken away from the parent.
Dandelion Seeds are blown away in the wind.
Sweet pea Seed pod explodes and seeds are distributed away from the parent.
Sycamore Seeds are blown away from the parent as they fall slowly through the
air.
Tomato Fruit eaten by animals, seeds are not digested and leave animal in
faeces.

Chapter Five

Differences in cartoons: changed head; straw; beetle; patch on sleeve; mouse; knot
in; label on bottle, top shelf; lid on jar, second shelf; missing bubble, extreme left;
extra spider's leg.

Similarities between two nymphs: one pair of antennae; six legs; separate head,
thorax and abdomen, etc..

1

A	B
6 legs	10 legs
wings	no wings
no claws	claws
separate parts	no separate parts

2

C	D
no legs	legs
no antennae	antennae
clitellum/saddle	no clitellum/saddle

3 (a) Movement   (d) Excretion
  (b) Response   (e) Feeding
  (c) Reproduction   (f) Energy need

4 Vertebrates: fish; frog; monkey; pigeon
  Invertebrates: earthworm; jellyfish

5 Warm-blooded: penguin; mouse
  Cold-blooded: turtle; lizard; snake; fish

6 M Reptile; **N** Mammal; **O** Fish; **P** Bird **Q** Amphibian

7 A Oak; **B** Lilac; **C** Fraxinus; **D** Platenus; **E** Alnus; **F** Aesculus

8

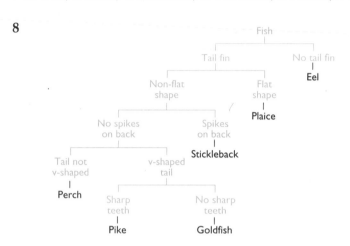

9 A Insect; **B** Myriapod; **C** Crustacean; **D** Arachnid

10 The information suggests that height is largely inherited since there is little difference between twins growing up together or apart. Weight seems to depend more on the environment in which they were brought up.

11 (a) Six legs; two antennae
  (b) Length of beetle; length of antennae; pattern on back
  (c) **Discontinuous variation** The beetles can be grouped according to their markings: no markings; spots; thin stripes; thick stripes.
    Continuous variation Length of the body, length of the antennae. You should try to measure these.
  (d) Your way of recording the information may vary.

Chapter Six

1 They have flattened bodies; claws to hold onto stones; and gills to enable them to get oxygen from the water.

2 (a) Aphids
  (b)

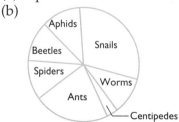

  (c) Centipedes occurred in fewer numbers; therefore, it should be higher up the pyramid of numbers.
  (d) Dampness; so they do not dry out. Lack of light.

3 Oak leaves → aphids → blue tits → kestrels

4 (a) Plant plankton

(b) Sand eel

(c) Animal plankton

(d) The food chain shows that the sand eel eats animals, but it does not show if it also eats plants.

(e) The Sun

5 (a) Foliage and bark → beetles → woodpeckers → kestrels

(b) Four

(c) Squirrel

6 (a) By rainwater draining through the soil into the rivers.

(b) The concentration of DDT in the fish would be too low to poison them.

(c) The DDT built up in the bodies of the otters. It took some time to reach a lethal level.

Chapter Seven

Solids at room temperature: sodium; iodine; sulphur; zinc; potassium chloride; sodium chloride; copper; iron

Liquids at room temperature: ethanol, mercury , bromine

Gases at room temperature: hydrogen, nitrogen, oxygen, ammonia

1 At –60°C mercury would be a solid and ammonia a liquid.

2 Temperature of the planet must be below –182°C so both sulphur dioxide and methane can be solid.

3 (a) (i) (ii)

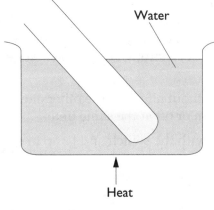

Water

Heat

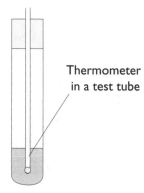

Thermometer in a test tube

(b) 41°C or 42°C   (c) 40°C

4 Water dissolves a greater range of solutes. Substances insoluble in water may be soluble in meths and vice versa.

Chapter Eight

1 The mass of the solution should be 110 g, i.e. the sum of the masses of solute and solvent.

The mass of water produced should be 10 g.

In both of these physical changes there should be no mass change.

2	Natural	Manufactured
	wood	bricks
	stone	glass
		plastic
		lead
		concrete
		tarmac
		roof tiles

Chapter Nine

1 A acid   **B** alkali   **C** neutral

2 (a) It would be impossible to see the colour because of the strong colour of blackcurrant cordial.
  (b) Bleach removes the colour from Universal indicator.

3 (a) zinc + hydrochloric acid → **zinc chloride** + **hydrogen**
  (b) sodium hydroxide + **hydrochloric acid** → sodium chloride + **water**
  (c) **calcium carbonate** + hydrochloric acid → calcium chloride + **water** + **carbon dioxide**

  (d) **magnesium** + **sulphuric acid** → magnesium sulphate + **hydrogen**

Chapter Ten

1 (a) Electrons are transferred from you onto the carpet.
  (b) You get a positive charge. You have lost electrons, so you have more positive charges than negative ones.

2 Each lamp can be switched on and off without affecting the other one. In the left-hand circuit, both lamps have to be on or off at the same time.

3 Your design should be a parallel circuit, with the switch and lamp in one branch and the variable resistor and motor in the other.

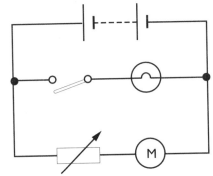

4 The motor, variable resistor, lamp and power supply or battery should all be in series.

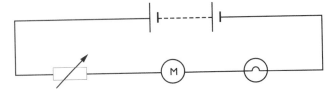

An alternative answer is to have the motor and lamp in parallel with each other and in series with the variable resistor.

5 (a) The lamp circuit has the greatest resistance. You can tell this because, for the same voltage, it passes the least current.
   (b) The current could be increased by using a higher voltage.
   (c) The resistance has decreased.

6 (a) The steel bar is magnetized; the end shown is a north-seeking pole.

   (b) The pattern would be similar, but the arrows would point in the opposite direction. This would happen because the opposite end must be a south-seeking pole.

7 The correct order is: E, B, A, C, D.

Chapter Eleven

1 The Earth pulls the apple.
   The finger pushes the switch.
   The dog pulls the boy.

2 (a) The resistive force
   (b) The driving force
   (c) The resistive and driving forces are equal.
   (d) Your diagram should show the driving force bigger than the resistive force.
   (e) Your diagram should show only one force: the resistive force.

3 (a) The speed is increasing, but the rate of increase gets less.
   (b) B and D. You can tell because the parachutist is travelling at a constant speed.
   (c) A. That is when she is speeding up.
   (d) The speed gets less. This is because the resistive force from the parachute is bigger than the weight.

4 (a) 50 m/s   (c) 625 km   (e) 15 m/s   (g) 36 m
   (b) 4 cm/s   (d) 3.5 hours   (f) 60 s

5 (a) 45 m   (c) 67.5 m
   (b) 4.5 s   (d) 112.5 m

6 (a) When a wire is used, the force acts over a smaller area, creating a bigger pressure.
   (b) If people wear high-heeled shoes, the weight presses down over a small area. The large pressure created makes dents in the bowling green surface.
   (c) The soft cushion moulds itself to your own shape, so your weight is spread over a larger area than when you sit on a hard stool.

7 (a) 12.5 N/m^2 or Pa   (d) 1000 N/m^2 or Pa
   (b) 250 N   (e) 300 000 N
   (c) 0.1 m^2

8 **Door handle** The pivot is at the centre of the handle and the force should be marked downwards on the handle, away from the pivot.
   Wheelbarrow The pivot is the point where the wheel touches the ground and the force should be marked upwards, acting on the handle.
   Spanner The pivot is the centre of the nut and the force should be horizontal, on the spanner handle.

9 180 Nm

10 **A** rotate anticlockwise   **D** rotate anticlockwise
   B rotate anticlockwise   **E** 5 N
   C balanced   **F** 10 N

Chapter Twelve

1 (a) C   (b) A and E   (c) B and D

2 Examples include a torch, a candle, a firework, a camping light, a glow-worm, a match, a fire, a gas hob, a hot wire.

3 Light from the lamp is reflected by the book into your eye.

4 D

5 (a) The lamp looks to be closer to the glass block than it really is.
  (b) Light travels in straight lines.

6 Yellow: red and green
  Cyan: green and blue
  Magenta: red and blue

7 (a) Red
  (b) (i)   Green
      (ii) Black
      (iii) Blue
      (iv) Blue

8 Green

9 (a) The rubber band vibrates.
  (b) The air molecules vibrate.
  (c) It sounds louder.
      The amplitude of vibration has increased.
  (d) The frequency of the vibrations has increased.

10 10 Hz and 25 000 Hz

11 The loud sounds cause excessive wear of the ossicles.

Chapter Thirteen

1 (a) The line is a curve, from the horizon on the left, through the position of the Sun shown, to the horizon on the right.
  (b) Your **S** should be higher than the position of the Sun shown in the diagram.
  (c) The Earth's rotation on its own axis.

2 (a) The densest planets are the innermost planets, closest to the Sun.
  (b) Pluto
  (c) The greatest mass
  (d) It has the lowest density or the smallest gravitational pull or it is the only one that is less massive than the Earth.

3 (a) The arrow should point from Pluto to the Sun.
  (b) It becomes greater
  (c) Pluto's speed increases

4 Satellites were used to take pictures of the troops and equipment. These pictures were transmitted back to Earth.

Chapter Fourteen

1
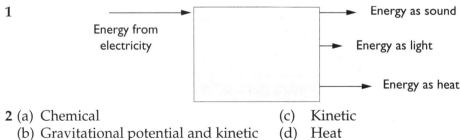

2 (a) Chemical   (c) Kinetic
  (b) Gravitational potential and kinetic   (d) Heat

3 (a) (i)   Sound
      (ii) Light

    (iii) Heat and kinetic
    (iv) Light and sound
    (v) Heat
  (b) Energy from fuel into gravitational potential energy.

4 2 J as sound

5 Renewable Solar power, wind power, timber, food, wave power, tidal power, hydroelectric power
Non-renewable Nuclear power, uranium ore, geothermal power, coal, natural gas, oil

6 (a) The Sun
  (b) Renewable; it relies on food that can be grown each year.
  (c) Chemical energy into heat energy.
  (d) It ends up as heat energy in the surroundings.

A NSWERS TO PRACTICE NATIONAL TEST QUESTIONS

Life processes and living things

1 (a) (i)

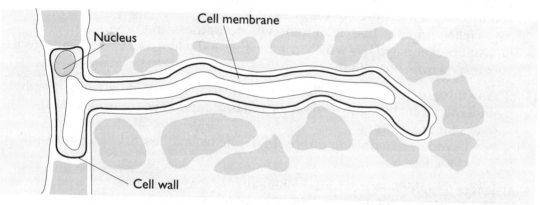

    (ii) The large surface area enables water to be absorbed from different parts of the soil and not just the part close to the plant.
  (b) Liquid inside the cell expands on freezing. This permanently damages the cell wall and the cell membrane.

2 (a) Smoking more cigarettes increases the risk of death from lung cancer.
  (b) From the graph, 10 cigarettes a day increases the risk of death from lung cancer by a factor of ten.

3 (a) (i) Stomach; liver   (b) (i)   Heart
    (ii) Heart; lungs   (ii) Lungs
                         (iii) Stomach

4 (a) To produce food (glucose, stored as starch) by photosynthesis using carbon dioxide and water in the presence of sunlight.
  (b) To absorb water and minerals from the soil. Also to anchor the plant.
  (c) To protect the flower in bud.
  (d) To produce pollen.

5 Ensures all plants are the same, i.e. like the parent plant.
New plants are produced quickly.
Avoids the high costs of seed.

6 (a) **A** fruit; **B** leaf; **C** stem; **D** root.
  (b) (i) Fruit   (ii) Root
  (c) (i) Ensure fertilization occurs, i.e. transfers pollen to the stigma of the flower.

(ii) Each flower will then produce a tomato fruit.

(iii) Flowers are then fully open.

7 (a) **A**

(b) Water is lost through the leaves.

(c) To ensure loss of mass is not due to evaporation of water in the test tube.

8 Breathe, grow, excrete, feed.

9 A Knotted wrack   **D** Tangleweed

   B Saw wrack   **E** Oarweed

   C Ladderwrack

10 (a) (i) Testes

      (ii) Ovary

(b) Uterus

(c) One egg cell fertilized. It then divided and each cell developed into a separate embryo.

(d) Two separate egg cells fertilized. Each developed into an embryo.

(e) Different diets, different environmental conditions

11 (a) Chorthippus   (b) 6 legs

12 Bluebells grow and flower before leaves are produced on the oak trees. The leaves would reduce photosynthesis by cutting off light.

Food for the plant is stored in the bulb.

Leaves fallen from the trees in autumn will decompose and produce humus.

The woodpecker has claws which grip the tree and a long beak which enables it to get insects from beneath the bark.

13 (a) Any three of the following: flat shape enables it to bury in the sand; mouth underneath enables it to feed on the sea bed; raised eyes or eyes on top; scales for protection.

(b) Fossils can be dated using carbon-14 dating techniques. Rocks are the same age as the fossils.

14 (a) leaves → aphids → ladybirds → blue tits → sparrowhawks

(b) The number of ladybirds will increase because of the greater food supply.

(c) The sparrowhawks can also eat blue tits.

(d) The sparrowhawks eat more blue tits so there are fewer blue tits to eat the ladybirds.

15

Prey	Predator
fly	spider
zebra	lion
fish	heron
worm	frog

Materials and their properties

1 (a) Can be squashed

(b) Block of iron

(c)

Material	Solid, liquid or gas
block of iron	solid
water	liquid
air	gas

(d) The iron filings are made up of very small particles.

2 (a) (i) Liquid to gas
(ii) Gas to liquid
(b) 100°C
(c) Liquid collected is pale blue.
Steam is leaving the test tube uncondensed.

3 (a) Nickel; copper; zinc
(b) Aluminium; silver

4 (a) **A** and **C**. They burn to give a single product.
(b) It produces an acidic oxide on burning.

5 (a) Liquid because it can flow and cannot be squashed.
(b) Gas because it can be squashed and is not hard.

6 (a) Some crystals are white and some are black.
(b) Large crystals.
(c) Action of high temperature and high pressure.

7 **C** Quartz; **D** Pyrite

8 (a) Wind carrying particles of sand.
(b) **F**. This layer of rock is less eroded.

9 (a) Galena
(b) (i) Test with acid
(ii) Density
(iii) Hardness or colour
(c) There is a difference in density. Fluorite will float in a frothy detergent/water mixture where galena sinks.

10 There is a loss of mass when zinc carbonate is heated. The mass of zinc oxide is unchanged.

11

	Seen at negative electrode	Seen at positive electrode
sodium nitrate solution	**bubbles of colourless gas**	**bubbles of colourless gas**
copper(II) nitrate solution	**red brown solid**	bubbles of greenish gas

(*Note:* You are not asked to identify the products.)

12 (a)

	Iron chloride solution	Zinc chloride solution
magnesium	✔	✔
copper	✗	✗
iron	✗	✗
zinc	✔	✗

(b) Zinc + copper nitrate → zinc nitrate + copper
(c) (i) Brown solid formed.
(ii) Above copper but below the other three metals in the reactivity series.

13 (a) Neutralization   (c) Reduction
(b) Polymerization   (d) Oxidation

14 **P** is above **S** in the reactivity series. **Q** is above **P** and **S**. **R** is above **Q**. The order is therefore **R Q P S**.

15 (a) (i) Carbon dioxide
      (ii) Dilute sulphuric acid
   (b) All the acid had reacted (**NOT** the solution is saturated)
   (c) Filtering
   (d) Cool the solution rapidly, e.g. put the solution into a test tube and cool the test tube in cold water.
   (e) Decomposition

Physical processes

1 A1 0.25 A   A3 0.3 A
  A2 0.25 A   A4 0.8 A

2 (a) Increase the voltage or add another battery.
      Remove one lamp or use lamps with a lower resistance.
   (b) Add a variable resistor in series with the lamps.
   (c) The left-hand lamp becomes brighter and the right-hand lamp goes out.
   (d) The left-hand lamp is brighter because more current passes in the circuit due to the resistance having decreased.
      The right-hand lamp goes out because most of the current goes through the switch, which is a low-resistance path.

3 (a) Iron
   (b) It is easily magnetized and it loses its magnetism when the current in the coil is switched off.

4 (a) It stays still.
   (b) The upward force needs to be bigger than the weight.
   (c) The forces are equal in size.
   (d) The rocket slows down because of the downward force; there is no upward force.

5 (a) The friction force between the wheel rim and the rubber block.
   (b) Water reduces the friction force.
   (c) The rough rims give a greater friction force and ensure that there is enough friction to stop even in wet weather.

6 (a) 1000 Pa or N/m^2.
   (b) The pressure doubles to 2000 Pa.

7 250 N/cm^2 or 2 500 000 Pa

8 (a) 45 Nm
   (b) 90 N

9 Your drawing should show light from the car being reflected at the driver's mirror into her eye.

10 Light travels faster than sound.

11 (a) Red and green
   (b) The blue jumper would look black and the yellow jumper would look green.

12 The amplitude is increased but the frequency is the same.

13 (a) Your ✗ should be on the opposite side of the Sun, showing the Earth having completed half an orbit.
   (b) One day
   (c) Light from the Sun is reflected from Mars back to the Earth.
   (d) The distance will decrease and then increase when the Earth has 'overtaken' Mars. This happens because the Earth is travelling faster than Mars and it also has less distance to travel to move through the same angle round the Sun.
   (e) Mercury is the planet nearest to the Sun. Venus is between Mercury and Earth.

14 (a) The part of the Earth on the right of the diagram is in darkness.
   (b) Sunlight is reflected from the Moon back down to Earth.
   (c) The Earth is tilted with the northern hemisphere facing the Sun.
   (d) None

15 (a) Coal, oil and gas
   (b) Tides
   (c) Wind and tides

16 It is wasted as heat energy in the surroundings.

17 (a) A loudspeaker or bell or buzzer
   (b) A kettle or fire or immersion heater
   (c) A vacuum cleaner or any device with an electric motor
   (d) A lamp or a torch

18 (a) Gravitational potential energy
   (b) Gravitational potential energy is transferred to kinetic energy.
   (c) It has been transferred to heat energy in the snow and the skis.

Scientific investigations

As part of your work in school you will have to carry out scientific investigations. These may be called AT1 by your teacher. These investigations will be marked by your teacher and you will be awarded a level. Your investigations will not be sent to an examiner outside the school at the end of Key Stage 3. The level will, however, be reported to your parents at the end of Key Stage 3 and will form the basis of similar work you will be required to do at Key Stage 4 for GCSE coursework.

Your scientific investigations do not require you to discover things but to use your scientific knowledge, and knowledge you can obtain from books, CD-ROM, etc., to make a prediction. This prediction is then tested by experiments which you have to plan and carry out. These experiments will produce results which you must record. Finally you have to look at your results and see if they fit your original prediction. At this stage you should evaluate your results and suggest any limitations they have and any improvements which could be made to your experiment.

We will now look at these aspects of scientific investigations in more detail, by splitting them into three areas.

❶ Making predictions and planning your experiments:
- Be sure you understand the scientific basis of any investigation you are carrying out.
- Make sure you make a clear prediction and, if possible, give a reason why, based on your scientific knowledge. This could take the form of:
  I think that..   **because** ....
- Make a clear plan of what you are going to do. What apparatus are you going to use? What observations and measurements are you going to make?
- Plan out a table to record your observations or measurements.
- Have you made sure that you have carried out a **fair test**? For example, if you are carrying out an investigation to compare the temperature rise of different samples of water with different fuels, are you using the same mass (or volume) of water in each case?

❷ Obtaining the evidence:
- Carry out your experiments safely.
- Make all of the observations and measurements carefully.
- Repeat any observation or measurement if you have any doubt about it.
- Consider carrying out experiments more than once and averaging measurements.
- Record the observations or measurements in a table, if possible, as you carry out the experiment.

❸ Looking at the results, drawing conclusions and considering the strength of your evidence:
- Arrange your results in a suitable order and look for patterns.
- Draw suitable charts or graphs to display your results.
- Use the results to draw conclusions.
- Use your scientific knowledge to try to explain the conclusions.
- Look critically at the conclusions to see if they are fully supported by the results.
- Identify any results which do not seem to fit the pattern.
- Suggest improvements you would make if you carried out the experiment again.

The keys to success in scientific investigation are getting a good prediction and carrying out thorough evaluation. These are the aspects of your investigation which will be of particular interest to your teacher.

Most students enjoy this aspect of Science as they can use their scientific knowledge to make and test predictions in the same way a scientist does.

*E*xperimental and Investigative Science

At the start of Key Stage 3 the majority of pupils will have reached at least Level 4 in Science. By the end of Key Stage 3 most pupils should be within the range of Levels 4–7. Levels 5–6 are the target for 14-year-olds. Level 8 is the standard reached by very able pupils.

Use our checklist to assess the Level reached, by ticking the skills that have been mastered.

Level 4

- [ ] Recognize the need for fair tests.
- [ ] Describe, or show in the way the task is performed, how to vary one factor whilst keeping others the same.
- [ ] Where appropriate, make predictions.
- [ ] Select suitable equipment to use and make a series of observations and measurements that are adequate for the task.
- [ ] Present observations and measurements clearly, using tables and bar charts.
- [ ] Begin to plot points to form simple graphs and use these graphs to point out and interpret patterns or trends in data.
- [ ] Take account of these patterns when drawing conclusions, and begin to relate conclusions to scientific knowledge and understanding.

Level 5

- [ ] Identify the key factors to consider in contexts that involve only a few factors.
- [ ] Where appropriate, make predictions based on scientific knowledge and understanding.
- [ ] Select apparatus for a range of tasks and use it with care.
- [ ] Make a series of observations or measurements with precision appropriate to the task.
- [ ] Begin to repeat observations and measurements and to offer simple explanations for any differences encountered.
- [ ] Record observations and measurements systematically and present data as line graphs.
- [ ] Draw conclusions that are consistent with the evidence and begin to relate these to scientific knowledge and understanding.

Level 6

- [ ] Use scientific knowledge and understanding to identify the key factors to consider and, where appropriate, make predictions.
- [ ] Make observations and measure with precision a variety of quantities, using instruments with fine divisions.
- [ ] Make enough measurements and observations for the task.
- [ ] Choose scales for graphs that enable appropriate data to be shown effectively.
- [ ] Identify measurements and observations that do not fit the main pattern or trend shown.
- [ ] Draw conclusions that are consistent with the evidence and explain these using scientific knowledge and understanding.

Level 7

- [ ] Use scientific knowledge and understanding to identify the key factors in situations involving a range of factors and, where appropriate, to make predictions.
- [ ] Make systematic observations and measurements with precision using a wide range of apparatus.
- [ ] Identify when repeat measurements and observations are needed in order to obtain reliable data.
- [ ] Present qualitative observations clearly and concisely.
- [ ] Present data in graphs, where appropriate, and use lines of best fit.
- [ ] Draw conclusions that are consistent with the evidence and explain these using scientific knowledge and understanding.
- [ ] Begin to consider whether the data collected are sufficient for the conclusions drawn.

Level 8

- [ ] Recognize that tasks of different kinds require different strategies.
- [ ] Use scientific knowledge and understanding to select an appropriate strategy, identifying the key factors to be considered.
- [ ] Where appropriate, make predictions based on knowledge and understanding.
- [ ] Manipulate apparatus with precision and skill.
- [ ] Decide which observations are relevant in qualitative work and include suitable detail in records of these.
- [ ] Decide the level of precision needed in measurements and collect data enabling relationships between variables to be tested.
- [ ] Identify and begin to explain anomalous observations and measurements, allowing for these when drawing graphs.
- [ ] Use knowledge and understanding to draw conclusions from evidence.
- [ ] Consider graphs and tables of results critically and identify shortcomings in the data obtained.

Exceptional performance

- [ ] Recognize that tasks of different kinds require different strategies.
- [ ] Use scientific knowledge and understanding to select an appropriate strategy, identifying the key factors to be considered, and making use of information from a range of relevant sources.
- [ ] Use scientific knowledge and understanding to make quantitative predictions where it is appropriate to do so.
- [ ] Manipulate a wide range of apparatus with precision and skill.
- [ ] Make records of relevant observations, clearly conveying points of significance.
- [ ] Decide the level of precision needed in measurements and collect data that satisfy these requirements.
- [ ] Use data to test relationships between variables.
- [ ] Identify and explain anomalous observations and measurements, allowing for these when drawing graphs.
- [ ] Use scientific knowledge and understanding to interpret the salient features of graphs and to draw conclusions from evidence.
- [ ] Present final results to an appropriate degree of precision.
- [ ] Consider graphs and tables of results critically and give reasoned accounts of how additional data to test conclusions could be collected.

Glossary

acceleration Increase in velocity (or speed) per second.

accommodation Ability of the eye to focus automatically.

acid Substance that dissolves in water to form a solution with a pH of less than 7. An acid usually contains hydrogen which can be replaced by a metal to form a salt.

algae Simple plants which may be single-celled or much larger (e.g. seaweeds).

alkali Base that dissolves in water to form a solution with a pH above 7. Alkalis are neutralized by acids to form salts.

alloy Metal made by mixing two or more metals together (e.g. brass is an alloy of copper and zinc).

alternating current (a.c.) Current which keeps changing direction around an electrical circuit.

ammeter Instrument for measuring current.

amoeba Single-celled animal found in the mud at the bottom of ponds.

amphibians Animals with backbones which have moist skins without scales. They live in and out of water (e.g. frogs).

anhydrous Substance without water. Often used to describe substances which have lost water of crystallization.

anion Negatively charged ion which moves towards the anode during electrolysis (e.g. chloride ion, Cl^-).

anode Positively charged electrode in electrolysis.

anther Male part of a flower which produces pollen.

antibodies Chemicals made by the body to kill germs.

aqueous solution Solution made by dissolving a substance in water.

arachnids Animals with four pairs of jointed legs (e.g. spiders).

arteries Blood vessels which carry blood away from the heart.

arthropods Animals with no backbone, but with jointed legs.

asexual reproduction Reproduction that requires only one parent (e.g. taking a geranium cutting).

atom Smallest part of an element that can exist that has all the properties of that element.

atria Upper two chambers of the heart.

bacteria Microscopic organisms which may cause decay.

barometer Instrument for measuring atmospheric pressure.

base Substance which reacts with an acid to form a salt and water only (e.g. metal oxides).

boiling Liquid turns rapidly to a vapour at a fixed temperature called the boiling point.

bones Hard substance which makes up the skeleton. Bones are made up of protein fibres and mineral salts.

capillary Smallest blood vessels in the body which link the arteries and the veins.

carbohydrates Compounds of carbon, hydrogen and oxygen which contain twice as much hydrogen as oxygen (e.g. glucose $C_6H_{12}O_6$).

carnivore Animal which eats flesh (e.g. tiger).

carpels Female reproductive organs in a flower.

catalyst Substance which alters the rate of a chemical reaction but is not itself used up in the reaction.

cathode Negatively charged electrode in electrolysis.

cation Positively charged ion which moves towards the cathode in electrolysis (e.g. hydrogen ion, H^+).

cell membrane Thin skin which surrounds an animal cell. Plant cells have a cellulose cell wall that surrounds the cell membrane.

cells Tiny units which make up living matter.

charges Either positive or negative, they exert forces on each other: like charges repel; unlike charges attract.

chemical change Change which results in the formation of new substances.

chlorophyll Green pigment in plants which absorbs light energy to start the photosynthesis reaction.

cholesterol Fatty material found in some foods which can be deposited in the arteries.

chromatography Way of separating mixtures, especially of coloured substances, by letting them spread across a filter paper or through an absorbent powder.

ciliary muscles Muscles which control the shape of the eye lens and keep it in focus.

classification Sorting out of information into groups; usually referring to living things.

combination Atoms of different elements joining together to form a compound.

combustion Burning of a substance in oxygen.

community Group of organisms which live together in a habitat.

components Parts that go to make up an electrical circuit.

compound Substance formed by joining atoms of different elements together. The properties of a compound are different from the elements that make it up. The proportions of the different elements in a particular compound are fixed.

concave Lens or mirror which curves inwards (i.e. cave-like).

condensation Occurs when a vapour turns to a liquid on cooling. Heat is given out during the change. Condensation is the opposite of evaporation.

conductor Allows electricity to pass through it (electrical conductor) or heat to pass through it (heat conductor). Metals are good conductors of heat and electricity. Carbon (graphite) is a good electrical conductor but a poor heat conductor.

convex Lens or mirror which curves outwards.

cornea Tough, outer coating on the front of the eye.

corrosion Wearing away of the surface of a metal by chemical attack (e.g. rusting of iron and steel).

crest Top of a wave. The lowest point on a wave is called the **trough**.

critical angle Angle above which the light is all internally reflected and not refracted.

crustaceans Arthropods which have two pairs of antennae, from nine to fifteen pairs of legs and often a hard, chalky shell.

crystal A piece of a substance that has a definite regular shape. Crystals of the same substance have the same shape. Slow crystallization will produce larger crystals.

current Flow of electric charge.

cytoplasm Jelly-like material which makes up the contents of a cell.

deceleration Decrease in velocity (or speed) over a given time, sometimes called negative acceleration.

decomposers Bacteria and fungi which break down the bodies of dead plants and animals and the waste products of living organisms.

decomposition Chemical reaction that results in the breaking down of substances into simpler ones.

density Mass of a particular volume of a substance. It is expressed as kg/m^3 or g/cm^3.

detergents Cleaning agents of two main types: soaps and soapless detergents.

diffuse reflection Light is reflected in all directions from a surface.

diffusion Movement of particles from where they are highly concentrated to where they are less concentrated.

digestion Breaking down of large, insoluble food molecules into smaller, soluble molecules.

direct current (d.c.) Current that stays in the same direction around an electrical circuit.

dispersion Splitting of white light into the different colours of the spectrum.

dissolving Occurs when a substance is added to water and disappears from view when stirred. The substance is still there and can be recovered by evaporation.

distillation Way of purifying a liquid or obtaining the solvent from a solution. The liquid is vaporized and the vapour condensed to reform the liquid. The condensed liquid is called the distillate.

ductile *See* metal.

ecology Scientific study of the way living things relate to one another and their environment.

efficiency Measure of the amount of energy wasted by a machine or process.

electrode Conducting rod, or plate, which carries electricity into, or out of, an electrolyte during electrolysis.

electrolysis Decomposition of an electrolyte, either molten or in aqueous solution, using electricity. The electrolyte is usually an acid, base, alkali or salt.

electrostatics Study of electric charges.

element Single pure substance that cannot be split up into any simpler chemical.

endothermic reaction Reaction that takes in heat.

energy Quantity that enables objects to do something or make something happen. Energy can be kinetic energy (due to movement) or potential energy (due to position).

environment Surroundings in which animals and plants live.

enzyme Protein which acts as a biological catalyst.

equilibrium Position when a set of forces is completely balanced.

erosion Wearing away of materials (e.g. rocks).

evaporation Process by which a liquid changes to its vapour. This happens at a temperature below the liquid's boiling point but is fastest when it is boiling.

exothermic reaction Reaction that gives out heat (e.g. the burning of coal).

eyepiece lens Lens in an optical instrument, such as a microscope, closest to the eye of the operator. The lens closest to the object is called the **objective lens**.

far point Furthest point which can be seen clearly.

fermentation Enzymes in yeast convert glucose into ethanol and carbon dioxide.

fertilization Joining together of a male sex cell with a female sex cell.

filtrate Liquid that comes through the filter paper during filtration.

filtration (or filtering) Method of separating a solid from a liquid. The solid collects on the filter paper and the liquid runs through.

flammable Substance which catches fire easily (e.g. petrol).

food chain Series of organisms (producers and consumers), starting with a green plant, which eat and are eaten by each other.

food web Series of interconnected food chains.

force Pushes, pulls, or turns which are measured in newtons (N).

fossil Preserved remains of organisms which lived millions of years ago.

fractional distillation Method of separating a mixture of different liquids using differences in their boiling points.

freezing A liquid changes to a solid at the freezing point. A pure substance will have a definite freezing point.

frequency Number of events in one second.

fuel Substance that burns easily to produce heat and light. A **fossil fuel** is present in the Earth in limited amounts only, and cannot be readily replaced (e.g. coal, petroleum).

genus A group of similar species.

germination Beginning of growth of a seed.

germs Microbes which cause disease.

gravity Attractive force of a body for a nearby object. The more massive the body the stronger will be the gravitational force.

habitat Part of an environment where a community of organisms live (e.g. a pond).

herbivore Animal which only eats plants (e.g. cow).

hermaphrodite Living organism which develops both male and female reproductive organs (e.g. earthworm).

hormone Chemical which regulates certain processes in the body.

humus Broken down remains of plants and animals found in the soil.

hydrolysis Splitting up of a compound using water.

igneous Rocks that have cooled and solidified from molten rock (e.g. granite).

immiscible Two liquids that do not mix (e.g. oil and water).

immunity Ability of the body to resist infection by disease-causing organisms.

incident ray Light ray which travels *into* an optical instrument.

indicator Chemical that can distinguish between an alkali and an acid by changing colour (e.g. litmus is red in acids and blue in alkalis).

inertia Reluctance of a mass to move, or have its movement changed.

insecticides Poisonous chemicals which can kill insects. They can often affect the environment.

insoluble Describes a substance that will not dissolve in a particular solvent.
insulator Substance which does not conduct electricity (e.g. rubber or plastic).
internal reflection Reflection that takes place on the inside surface of the material.
ion Positively, or negatively, charged particle formed when an atom, or group of atoms, lose or gain electrons.
iris Coloured, circular portion at the front of the eye. It adjusts to alter the amount of light entering the eye.

key Series of written instructions used to identify living organisms.
kinetic energy *See* energy.

laterally inverted The image is turned the wrong way round.
lens An optical device made of glass, or plastic, which bends light.
long-sighted Person who is not able to see close objects clearly without correction. It is corrected with a convex lens.

malleable *See* metal.
mass The amount of matter in a material. It is important to distinguish mass from weight.
melt A solid changes to a liquid at the melting point.
metal An element that is: shiny; conducts heat and electricity; can be beaten into thin sheets (**malleable**); or drawn in wires (**ductile**); is probably a metal.
metamorphic Igneous or sedimentary rocks were thoroughly altered by heat and/or pressure within the crust of the Earth without melting to form these rocks (e.g. marble).
microbes Microscopic plants and animals including bacteria and viruses. Bacteria can be killed by antibiotics, such as penicillin, but viruses cannot.
mineral Naturally occurring substance from which rocks are made.
mixture Substance made by just mixing other substances together. This is not a chemical reaction.
molecule Smallest part of a diatomic element (e.g. H_2) or compound that can exist on its own.

near point Closest point which can be seen clearly.

nectar Sugary liquid produced by plants to attract insects for pollination.
neutralization A reaction where an acid is cancelled out by a base or alkali.

omnivore Animal which eats plants and animals (e.g. human).
organ Collection of tissues grouped together to make a structure with a specific job (e.g. the heart).
osmosis Movement of water from a dilute solution to a more concentrated solution through a semi-permeable membrane (i.e. movement from a high concentration of water to a low concentration of water).
ovaries Female sex organs which produce ova.
ovum Female sex cell, sometimes called an egg.
oxidation Reaction where a substance gains oxygen or loses hydrogen.

pH Measure of the acidity, or alkalinity, of a substance. A substance with a pH of 7 is neutral.
photosynthesis Process by which green plants use the energy from the sun to fuel the reaction that builds up their food.
plasma Liquid part of the blood.
pollution Substances in the environment which are harmful to living things.
potential energy *See* energy.
power Rate of energy transfer.
precipitate Insoluble substance formed in a chemical reaction which causes the solution to become cloudy.
predator Animal which catches and eats other animals (prey).
pressure Force per unit area.
properties Description of a substance and how it behaves. Physical properties include density and melting point. Chemical properties describe chemical changes.
proteins Important body-building foods including meat, fish and eggs.
pupil Hole through which light passes in the centre of the iris of an eye.

random motion Uncontrolled motion which cannot be predicted.
reactant Chemical substance which takes part in a chemical reaction.
real image Image which can be projected onto a screen.

red blood cells Contain haemoglobin and carry oxygen around the body via the blood system.

redox reaction Reaction where both *re*duction and *oxi*dation occur.

reduction Opposite of oxidation: a reaction where oxygen is lost and hydrogen is gained.

refraction Change of direction of light as it passes from one material to another.

reproduction Process by which living organisms produce more of their own species.

residue Insoluble substance left on the filter paper during filtration.

resistance Opposition to current in a circuit by a material, measured in ohms (Ω).

resonance Object vibration in sympathy with something else.

respiration Chemical process in living things which releases energy from food.

retina Light-sensitive coating on the back of the inside of the eye.

salt Substance which is formed as a product of a neutralization. A salt is the product when hydrogen in an acid is replaced by a metal.

scalar Measurement that only has size, not direction (e.g. speed).

sedimentary Rocks composed of compacted fragments of older rocks, and other minerals, which have accumulated on the floor of a sea or lake (e.g. sandstone).

sepals Leaf-like structures which enclose and surround an unopened flower bud.

sex cells Cells which join together during fertilization.

sex organs Parts of the body which produce sex cells.

short-sight Inability to see faraway objects clearly without correction. It is corrected using a concave lens.

soil Complex mixture of rock fragments, mineral salts, humus and living organisms in which most of the world's plants grow.

solubility Grams of a solute that will dissolve in 100 g of solvent at a particular temperature.

solute Substance that dissolves in a solvent to form a solution.

solvent Liquid in which the solute dissolves.

species Group of organisms which can breed with each other to produce fertile offspring.

stamens Male sex organs in flowering plants.

stigma Part of the carpel to which the pollen grains stick during pollination.

synthesis Formation of a compound from the elements that make it up.

tissues Groups of similar cells which perform a similar task.

toxin Poisonous substance, often produced by bacteria.

umbra Area of complete shadow. Partial shadow is called **penumbra**.

vacuole Fluid-filled space inside a cell.

vapour Gas that will condense to a liquid on cooling to room temperature.

vector Measurement that has both size and direction (e.g. velocity).

veins Vessels in animals which return blood to the heart.

virtual image Image which cannot be projected on to a screen.

volatile Describes a liquid which turns easily to a vapour (e.g. petrol).

voltage Indicates the energy value of electric charge. It is measured in volts (V).

water of crystallization Definite amount of water bound up in a crystal (e.g. hydrous copper(II) sulphate, $CuSO_4.5H_2O$).

wavelength Distance between two identical points on a wave.

white blood cells Blood cells which help the body fight disease. They do not contain haemoglobin.

work Work is done when a force moves.

yeast Single-celled fungus which releases carbon dioxide when it respires.

*I*ndex